Earth Science

Study Guide

HOLT, RINEHART AND WINSTON

A Harcourt Education Company

Orlando • **Austin** • New York • San Diego • Toronto • London

TO THE STUDENT

Do you need to practice for an upcoming section quiz or chapter test? If so, this booklet will help you. The *Study Guide* is a tool that allows you to confirm what you know and to identify topics you do not understand so that you can succeed in your studies. These worksheets are reproductions of each Section Review and Chapter Review in your textbook, but there is one difference—the *Study Guide* worksheets provide plenty of space for you to record your answers and write your thoughts and ideas.

Use these worksheets in the following ways:

- as a learning tool to work interactively with the textbook by answering the questions as you read the text
- as a review to test your understanding of the main concepts and terminology
- as a practice quiz or test to prepare for a section quiz or chapter test

ISBN-13: 978-0-03-099449-4
ISBN-10: 0-03-099449-7

3 4 5 6 7   054   10 09 08

Contents

The World of Earth Science

Section Review Worksheets

Chapter Review Worksheets

Maps as Models of the Earth

Section Review Worksheets

Chapter Review Worksheets

Minerals of the Earth's Crust

Section Review Worksheets

Chapter Review Worksheets

Rocks: Mineral Mixtures

Section Review Worksheets

Chapter Review Worksheets

Energy Resources

Section Review Worksheets

Chapter Review Worksheets

The Rock and Fossil Record

Plate Tectonics

Earthquakes

Volcanoes

Weathering and Soil Formation

The Flow of Fresh Water

Section Review Worksheets

Agents of Erosion and Deposition

Section Review Worksheets

Exploring the Oceans

Section Review Worksheets

The Movement of Ocean Water

Section Review Worksheets

The Atmosphere

Section Review Worksheets

Understanding Weather

Climate

Studying Space

Stars, Galaxies, and the Universe

Formation of the Solar System

A Family of Planets

Section Review Worksheets

Exploring Space

Section Review Worksheets

Section Review

Branches of Earth Science

USING KEY TERMS

1. Use each of the following terms in a separate sentence: *geology,*
   oceanography, and *astronomy.*

   _____

   _____

   _____

   _____

UNDERSTANDING KEY IDEAS

_____ 2. Which of the following Earth scientists would study tornadoes?
   a. a geologist   **c.** a meteorologist
   b. an oceanographer   **d.** an astronomer

_____ 3. On which major branch of Earth science does geochemistry rely?
   a. geology   **c.** meteorology
   b. oceanography   **d.** astronomy

4. List the major branches of Earth science.

   _____

   _____

   _____

   _____

5. In which major branch of Earth science would a scientist study black smokers?

   _____

   _____

6. List two branches of Earth science that rely heavily on other areas of science.
   Explain how the branches rely on the other areas of science.

   _____

   _____

   _____

   _____

Section Review *continued*

7. List and describe three Earth science careers.

_____

_____

_____

MATH SKILLS

8. Each week, a volcanologist reads 80 pages in a book about volcanoes. In a 4-week period, how many pages will the volcanologist read? Show your work below.

CRITICAL THINKING

9. Making Inferences If you were a *hydrogeologist*, what kind of work would you do?

_____

_____

_____

_____

10. Identifying Relationships Explain why an ecologist might need to understand geology.

_____

_____

11. Applying Concepts Explain how an airline pilot would use Earth science in his or her career.

_____

_____

_____

Section Review

Scientific Methods in Earth Science

USING KEY TERMS

1. Use the following terms in the same sentence: *scientific method* and *hypothesis*.

_____

_____

_____

UNDERSTANDING KEY IDEAS

_____ **2.** Which of the following is NOT part of scientific methods?
   a. ask a question
   b. test the hypothesis
   c. analyze results
   d. close the case

_____ **3.** Which of the following is the step in the scientific methods in which a scientist uses a controlled experiment?
   a. form a hypothesis
   b. test the hypothesis
   c. analyze results
   d. communicate results

4. Explain how scientists use more than imagination to form answers about the natural world.

_____

_____

_____

_____

5. Why do scientists communicate the results of their investigations?

_____

_____

_____

_____

6. For what reason might a scientist change his or her hypothesis after it has already been accepted?

_____

_____

_____

_____

_____

MATH SKILLS

7. If the *Seismosaurus's* neck is 20 m long and the scientist studying *Seismosaurus* is 2 m long, how many scientists, lined up head to toe, would it take to equal the length of a *Seismosaurus's* neck? Show your work below.

CRITICAL THINKING

8. Applying Concepts Why might two scientists develop different hypotheses based on the same observations? Explain.

_____

_____

_____

9. Evaluating Hypotheses Explain why Gillette's hypothesis—that the bones came from a kind of dinosaur unknown to science—is a testable hypothesis.

_____

_____

_____

_____

Section Review

Scientific Models

USING KEY TERMS

1. In your own words, write a definition for each of the following terms: *model* and *theory*.

_____

_____

UNDERSTANDING KEY IDEAS

_____ **2.** Which of the following types of models are systems of ideas?
   a. physical models
   b. mathematical models
   c. conceptual models
   d. climate models

3. Why do scientists use models?

_____

_____

_____

_____

4. Describe the three types of models.

_____

_____

_____

5. Which type of model would you use to study objects that are too small to be seen? Explain.

_____

_____

_____

_____

| Section Review *continued*

6. Describe why the climate model is a mathematical model.

_____

_____

_____

_____

MATH SKILLS

7. A model of a bridge is 1 m long and 2.5% of the actual size of the bridge. How long is the actual bridge? Show your work below.

CRITICAL THINKING

8. Analyzing Ideas Describe one advantage of physical models.

_____

_____

_____

_____

9. Applying Concepts What type of model would you use to study an earthquake? Explain.

_____

_____

_____

_____

Section Review

Measurement and Safety

USING KEY TERMS

The statements below are false. For each statement, replace the underlined term to make a true statement.

1. The length multiplied by the width of an object is the <u>density</u> of the object.

_____

2. The measure of the amount of matter in an object is the <u>area</u>.

_____

UNDERSTANDING KEY IDEAS

_____ **3.** Which of the following SI units is most often used to measure length?
   a. meter
   b. liter
   c. gram
   d. degrees Celsius

4. What are two benefits of using the International System of Units?

_____

_____

5. At what temperature in degrees Celsius does water freeze?

_____

_____

6. Why is it important to follow safety rules?

_____

_____

_____

_____

| Section Review *continued*

MATH SKILLS

7. Find the density of an object that has a mass of 34 g and a volume of 14 mL. Show your work below.

CRITICAL THINKING

8. Making Comparisons Which weighs more: a pound of feathers or a pound of lead? Explain.

_____

_____

_____

_____

Skills Worksheet

Chapter Review

USING KEY TERMS

Complete each of the following sentences by choosing the correct term from the word bank.

geology

scientific methods

astronomy

hypothesis

1. The study of the origin, history, and structure of the Earth and the processes that shape the Earth is called _____.

2. An explanation that is based on prior scientific research or observations and that can be tested is called a(n) _____.

3. _____ are a series of steps followed to solve problems.

UNDERSTANDING KEY IDEAS

Multiple Choice

_____ **4.** The science that uses geology to study how humans affect the natural environment is
   a. paleontology.
   b. environmental science.
   c. cartography.
   d. volcanology.

_____ **5.** A pencil measures 14 cm long. How many millimeters long is it?
   a. 1.4 mm
   b. 140 mm
   c. 1,400 mm
   d. 1,400,000 mm

_____ **6.** Which of the following is NOT an SI unit?
   a. meter
   b. foot
   c. liter
   d. degrees Celsius

_____ **7.** Which of the following is a limitation of models?
   a. They are large enough to be seen.
   b. They do not act exactly like the thing they model.
   c. They are smaller than the thing they model.
   d. They use familiar things to model unfamiliar things.

_____ **8.** Gillette's hypothesis was
   a. supported by his results.
   b. not supported by his results.
   c. based only on observations.
   d. based only on what he already knew.

| Chapter Review *continued*

Short Answer

9. Why would scientific investigations lead to new scientific investigations?

_____

_____

_____

_____

10. How and why do scientists use models?

_____

_____

_____

_____

11. What are three types of models? Give an example of each.

_____

_____

_____

_____

12. What problems could occur if scientists didn't communicate the results of their investigations?

_____

_____

_____

_____

13. What problems could occur if there were not an International System of Units?

_____

_____

_____

_____

14. Which safety symbols would you expect to see for an experiment that requires the use of acid?

_____

_____

_____

_____

CRITICAL THINKING

15. Concept Mapping Use the following terms to create a concept map: *Earth science, scientific methods, hypothesis, problem, question, experiment,* and *observations.*

Chapter Review *continued*

16. Analyzing Processes Why do you not need to complete the steps of scientific methods in a specific order?

_____

_____

_____

_____

_____

17. Evaluating Conclusions Why might two scientists working on the same problem draw different conclusions?

_____

_____

_____

18. Analyzing Methods Scientific methods often begin with observation. How does observation limit what scientists can study?

_____

_____

_____

_____

19. Making Comparisons A rock that contains fossil seashells might be studied by scientists in at least two branches of Earth science. Name those branches. Why did you choose those two branches?

_____

_____

_____

_____

❚ Chapter Review *continued*

INTERPRETING GRAPHICS

Use the graph below to answer the questions that follow.

Atmospheric CO$_2$ (1860–1980)

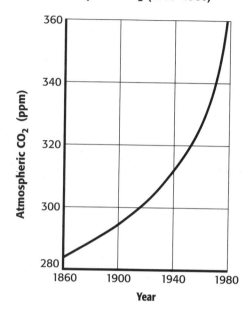

20. Has the amount of CO$_2$ in the atmosphere increased or decreased since 1860?

_____

_____

21. The line on the graph is curved. What does this curve indicate?

_____

_____

_____

22. Was the rate of change in the level of CO$_2$ between 1940 and 1960 higher or lower than it was between 1880 and 1900? How can you tell?

_____

_____

_____

23. What conclusions can you draw from reading this graph?

_____

_____

_____

Skills Worksheet

Section Review

You Are Here
USING KEY TERMS

1. Use each of the following terms in a separate sentence: *latitude*, *longitude*, *equator*, and *prime meridian*.

_____

_____

_____

_____

2. In your own words, write a definition for the term *true north*.

_____

_____

UNDERSTANDING KEY IDEAS

_____ **3.** The geographic poles are
   a. used as reference points when describing direction and location on Earth.
   b. formed because of the Earth's magnetic field.
   c. at either end of the Earth's axis.
   d. Both (a) and (c)

4. How are lines of latitude and lines of longitude alike? How are they different?

_____

_____

5. How can you use a magnetic compass to find directions on Earth?

_____

_____

6. What is the difference between true north and magnetic north?

_____

_____

Section Review *continued*

7. How do lines of latitude and longitude help you find locations on the Earth's surface?

_____

_____

MATH SKILLS

8. The distance between 40° N latitude and 41° N latitude is 69 mi. What is this distance in km? Show your work below. (Hint: 1 km = 0.621 mi)

CRITICAL THINKING

9. Applying Concepts While exploring the attic, you find a treasure map. The map shows that the treasure is buried at 97° north and 188° east. Explain why this location is incorrect.

_____

_____

_____

10. Making Inferences When using a compass to map an area, why do you need to know an area's magnetic declination?

_____

_____

_____

_____

Skills Worksheet

Section Review

Mapping the Earth's Surface

USING KEY TERMS

1. In your own words, write a definition for each of the following terms: *cylindrical projection*, *azimuthal projection*, and *conic projection*.

_____

_____

_____

UNDERSTANDING KEY IDEAS

_____ 2. Which of the following map projections is most often used to map the United States?
   a. cylindrical projection
   b. conic projection
   c. azimuthal projection
   d. equal-area projection

3. List five things found on maps. Explain how each thing is important to reading a map.

_____

_____

_____

_____

_____

4. Describe how GPS can help you find your location on Earth.

_____

_____

5. Why is radar useful when mapping areas that tend to be covered in clouds?

_____

_____

CRITICAL THINKING

6. Analyzing Ideas Imagine you are a mapmaker. You have been asked to map a landmass that has more area from east to west than from north to south. What type of map projection would you use? Explain.

_____

_____

_____

7. Making Inferences Imagine looking at a map of North America. Would this map have a large scale or a small scale? Would a map of your city have a large scale or a small scale? Explain.

_____

_____

_____

INTERPRETING GRAPHICS

Use the map below to answer the questions that follow.

8. What type of projection was used to make this map?

_____

9. Which areas of this map are the most distorted? Explain.

_____

_____

10. Which areas of this map are the least distorted? Explain.

_____

_____

Skills Worksheet

Section Review

Topographic Maps

USING KEY TERMS

1. In your own words, write a definition for each of the following terms: *topographic map, contour interval,* and *relief.*

_____

_____

_____

UNDERSTANDING KEY IDEAS

_____ **2.** An index contour
     a. is a heavier contour line that shows a change in elevation.
     b. points in the direction of higher elevation.
     c. indicates a depression.
     d. indicates a hill.

3. How do topographic maps represent the Earth's surface?

_____

_____

_____

4. How does the relief of an area determine the contour interval used on a map?

_____

_____

_____

5. What are the rules of contour lines?

_____

_____

_____

_____

MATH SKILLS

6. The contour line at the base of a hill reads 90 ft. There are five contour lines between the base of the hill and the top of the hill. If the contour interval is 30 ft, what is the elevation of the highest contour line? Show your work below.

CRITICAL THINKING

7. Making Inferences Why isn't the highest point on a hill represented by a contour line?

_____

_____

_____

_____

Skills Worksheet)

Chapter Review

USING KEY TERMS

For each pair of terms, explain how the meanings of the terms differ.

1. *true north* and *magnetic north*

_____

_____

2. *latitude* and *longitude*

_____

_____

3. *equator* and *prime meridian*

_____

_____

4. *cylindrical projection* and *azimuthal projection*

_____

_____

5. *contour interval* and *index contour*

_____

_____

6. *global positioning system* and *geographic information system*

_____

_____

UNDERSTANDING KEY IDEAS

Multiple Choice

_____ **7.** A point whose latitude is 0° is located on the
   a. North Pole.   **c.** South Pole.
   b. equator.   **d.** prime meridian.

_____ **8.** The distance in degrees east or west of the prime meridian is
   a. latitude.   **c.** longitude.
   b. declination.   **d.** projection.

Chapter Review continued

_____ 9. Widely spaced contour lines indicate a
    a. steep slope.   **c.** hill.
    b. gentle slope.   **d.** river.

_____ 10. The most common map projections are based on three geometric shapes. Which of the following geometric shapes is NOT one of the three geometric shapes?
    a. cylinder   **c.** cone
    b. square   **d.** plane

_____ 11. A cylindrical projection is distorted near the
    a. equator.   **c.** prime meridian.
    b. poles.   **d.** date line.

_____ 12. What is the relationship between the distance on a map and the actual distance on Earth called?
    a. legend   **c.** relief
    b. elevation   **d.** scale

_____ 13. _____ is the height of an object above sea level.
    a. Contour interval   **c.** Declination
    b. Elevation   **d.** Index contour

Short Answer

14. List four methods that modern mapmakers use to make accurate maps.

_____

_____

_____

_____

15. Why is a map legend important?

_____

_____

16. Why does Greenland appear so large in relation to other landmasses on a map made using a cylindrical projection?

_____

_____

17. What is the function of contour lines on a topographic map?

_____

_____

18. How can GPS help you find your location on Earth?

_____

_____

19. What is GIS?

_____

_____

CRITICAL THINKING

20. Concept Mapping Use the following terms to create a concept map: *maps, legend, map projection, map parts, scale, cylinder, title, cone, plane, date,* and *compass rose.*

21. Making Inferences One of the important parts of a map is its date. Why is the date important?

_____

_____

_____

_____

22. Analyzing Ideas Why is it important for maps to have scales?

_____

_____

_____

_____

23. Applying Concepts Imagine that you are looking at a topographic map of the Grand Canyon. Would the contour lines be spaced close together or far apart? Explain your answer.

_____

_____

_____

_____

24. Analyzing Processes How would a GIS system help a team of engineers plan a new highway system for a city?

_____

_____

_____

_____

25. Making Inferences If you were stranded in a national park, what kind of map of the park would you want to have with you? Explain your answer.

_____

_____

_____

_____

INTERPRETING GRAPHICS

Use the topographic map below to answer the questions that follow.

26. What is the elevation change between two adjacent lines on this map?

_____

_____

27. What type of relief does this area have?

_____

_____

28. What surface features are shown on this map?

_____

_____

29. What is the elevation at the top of Ore Hill?

_____

_____

Skills Worksheet

Section Review

What Is a Mineral?

USING KEY TERMS

1. In your own words, write a definition for each of the following terms: *element, compound,* and *mineral.*

_____

_____

_____

UNDERSTANDING KEY IDEAS

_____ 2. Which of the following minerals is a nonsilicate mineral?
   a. mica
   b. quartz
   c. gypsum
   d. feldspar

3. What is a crystal, and what determines a crystal's shape?

_____

_____

4. Describe the two major groups of minerals.

_____

_____

MATH SKILLS

5. If there are approximately 3,600 known minerals and about 20 of the minerals are native elements, what percentage of all minerals are native elements? Show your work below.

| Section Review *continued*

CRITICAL THINKING

6. Applying Concepts Explain why each of the following is not considered a mineral: water, oxygen, honey, and teeth.

_____

_____

_____

_____

7. Applying Concepts Explain why scientists consider ice to be a mineral.

_____

_____

_____

_____

8. Making Comparisons In what ways are sulfate and sulfide minerals the same. In what ways are they different?

_____

_____

_____

_____

Skills Worksheet

Section Review

Identifying Minerals
USING KEY TERMS

1. Use each of the following terms in a separate sentence: *luster*, *streak*, and *cleavage*.

_____

_____

_____

UNDERSTANDING KEY IDEAS

_____ **2.** Which of the following properties of minerals is expressed in numbers?
    a. fracture
    b. cleavage
    c. hardness
    d. streak

3. How do you determine a mineral's streak?

_____

_____

4. Briefly describe the special properties of minerals.

_____

_____

MATH SKILLS

5. If a mineral has a specific gravity of 5.5, how much more matter is there in 1 cm³ of this mineral than in 1 cm³ of water? Show your work below.

Section Review *continued*

CRITICAL THINKING

6. Applying Concepts What properties would you use to determine whether two mineral samples are different minerals?

_____

_____

_____

_____

7. Applying Concepts If a mineral scratches calcite but is scratched by apatite, what is the mineral's hardness?

_____

_____

_____

_____

8. Analyzing Methods What would be the easiest way to identify calcite?

_____

_____

_____

_____

Skills Worksheet

Section Review

The Formation, Mining, and Use of Minerals

USING KEY TERMS

Complete each of the following sentences by choosing the correct term from the word bank.

ore   reclamation

1. _____ is the process of returning land to its original condition after mining is completed.

2. _____ is the term used to describe a mineral deposit that is large enough and pure enough to be mined for profit.

UNDERSTANDING KEY IDEAS

_____ 3. Which of the following conditions is NOT important in the formation of minerals?
   a. presence of groundwater
   b. evaporation
   c. volcanic activity
   d. wind

4. What are the two main types of mining, and how do they differ?

   _____

   _____

5. List some uses of metallic minerals.

   _____

   _____

6. List some uses of nonmetallic minerals.

   _____

   _____

MATH SKILLS

7. A diamond cutter has a raw diamond that weighs 19.5 carats and from which two 5-carat diamonds will be cut. How much did the raw diamond weigh in milligrams? How much will each of the two cut diamonds weigh in milligrams? Show your work below.

CRITICAL THINKING

8. Analyzing Ideas How does reclamation protect the environment around a mine?

_____

_____

_____

_____

9. Applying Concepts Suppose you find a mineral crystal that is as tall as you are. What kinds of environmental factors would cause such a crystal to form?

_____

_____

_____

_____

Skills Worksheet

Chapter Review

USING KEY TERMS

1. Use each of the following terms in a separate sentence: *element, compound,* and *mineral.*

_____

_____

_____

_____

For each pair of terms, explain how the meanings of the terms differ.

2. *color* and *streak*

_____

_____

3. *mineral* and *ore*

_____

_____

4. *silicate mineral* and *nonsilicate mineral*

_____

_____

UNDERSTANDING KEY IDEAS

Multiple Choice

_____ **5.** Which of the following properties of minerals does Mohs scale measure?
   a. luster
   b. hardness
   c. density
   d. streak

_____ **6.** Pure substances that cannot be broken down into simpler substances by ordinary chemical means are called
   a. molecules.
   b. elements.
   c. compounds.
   d. crystals.

_____ **7.** Which of the following properties is considered a special property that applies to only a few minerals?
   a. luster
   b. hardness
   c. taste
   d. density

_____ **8.** Silicate minerals contain a combination of the elements

    a. sulfur and oxygen.   **c.** iron and oxygen.

    b. carbon and oxygen.   **d.** silicon and oxygen.

_____ **9.** The process by which land used for mining is returned to its original state is called

    a. recycling.   **c.** reclamation.

    b. regeneration.   **d.** renovation.

_____ **10.** Which of the following minerals is an example of a gemstone?

    a. mica   **c.** gypsum

    b. diamond   **d.** copper

SHORT ANSWER

11. Compare surface and subsurface mining.

_____

12. Explain the four characteristics of a mineral.

_____

_____

_____

_____

13. Describe two environments in which minerals form.

_____

_____

14. List two uses for metallic minerals and two uses for nonmetallic minerals.

_____

_____

15. Describe two ways to reduce the effects of mining.

_____

_____

16. Describe three special properties of minerals.

_____

_____

_____

CRITICAL THINKING

17. Concept Mapping Use the following terms to create a concept map: *minerals, calcite, silicate minerals, gypsum, carbonates, nonsilicate minerals, quartz,* and *sulfates.*

18. Making Inferences Imagine that you are trying to determine the identity of a mineral. You decide to do a streak test. You rub the mineral across the streak plate, but the mineral does not leave a streak. Has your test failed? Explain your answer.

_____

_____

_____

_____

19. Applying Concepts Why would cleavage be important to gem cutters, who cut and shape gemstones?

_____

_____

_____

_____

20. Applying Concepts Imagine that you work at a jeweler's shop and someone brings in some gold nuggets for sale. You are not sure if the nuggets are real gold. Which identification tests would help you decide whether the nuggets are gold?

_____

_____

_____

_____

21. Identifying Relationships Suppose you are in a desert. You are walking across the floor of a dry lake, and you see crusts of cubic halite crystals. How do you suppose the halite crystals formed? Explain your answer.

_____

_____

_____

_____

_____

Chapter Review *continued*

INTERPRETING GRAPHICS

The table below shows the temperatures at which various minerals melt. Use the table below to answer the questions that follow.

Melting Points of Various Minerals	
Mineral	**Melting Point (°C)**
Mercury	−39
Sulfur	+113
Halite	801
Silver	916
Gold	1,062
Copper	1,083
Pyrite	1,171
Fluorite	1,360
Quartz	1,710
Zircon	2,500

22. According to the table, what is the approximate difference in temperature between the melting points of the mineral that has the lowest melting point and the mineral that has the highest melting point?

_____

23. Which of the minerals listed in the table do you think is a liquid at room temperature?

_____

_____

24. Pyrite is often called *fool's gold*. Using the information in the table, how could you determine if a mineral sample is pyrite or gold?

_____

_____

25. Convert the melting points of the minerals shown in the table from degrees Celsius to degrees Fahrenheit. Use the formula °F = (9/5 × °C) + 32.

_____

_____

Skills Worksheet

Section Review

The Rock Cycle

USING KEY TERMS

Complete each of the following sentences by choosing the correct term from the word bank.

rock   composition

rock cycle   texture

1. The minerals that a rock is made of determine the _____

of that rock.

2. _____ is a naturally occurring solid mixture of one or

more minerals.

UNDERSTANDING KEY IDEAS

_____ **3.** Sediments are transported or moved from their original source by a
process called

a. deposition.

b. erosion.

c. uplift.

d. weathering.

4. Describe two ways that rocks have been used by humans.

_____

_____

5. Name four processes that change rock inside the Earth.

_____

_____

_____

_____

6. Describe four processes that shape Earth's surface.

_____

_____

_____

_____

7. Give an example of how texture can provide clues as to how and where a rock formed.

_____

_____

CRITICAL THINKING

8. Making Comparisons Explain the difference between texture and composition.

_____

_____

_____

_____

9. Analyzing Processes Explain how rock is continually recycled in the rock cycle.

_____

_____

_____

_____

INTERPRETING GRAPHICS

10. Look at the table below. Sandstone is a type of sedimentary rock. If you had a sample of sandstone that had an average particle size of 2 mm, what texture would your sandstone have?

Classification of Clastic Sedimentary Rocks	
Texture	Particle Size
coarse grained	> 2 mm
medium grained	0.06 to 2 mm
fine grained	< 0.06 mm

Section Review

Igneous Rock

USING KEY TERMS

1. In your own words, write a definition for each of the following terms: *intrusive igneous rock* and *extrusive igneous rock*.

_____

_____

_____

_____

UNDERSTANDING KEY IDEAS

2. _____ is an example of a coarse-grained, felsic, igneous rock.
    a. Basalt
    b. Gabbro
    c. Granite
    d. Rhyolite

3. Explain three ways in which magma can form.

_____

_____

_____

4. What determines the texture of igneous rocks?

_____

_____

_____

_____

| Section Review *continued*

MATH SKILLS

5. The summit of a granite batholith has an elevation of 1,825 ft. What is the height of the batholith in meters? Show your work below.

CRITICAL THINKING

6. Making Comparisons Dikes and sills are both types of igneous intrusive bodies. What is the difference between a dike and a sill?

_____

_____

_____

_____

7. Predicting Consequences An igneous rock forms from slow-cooling magma deep beneath the surface of the Earth. What type of texture is this rock most likely to have? Explain.

_____

_____

_____

_____

Skills Worksheet

Section Review

Sedimentary Rock
USING KEY TERMS

1. In your own words, write a definition for each of the following terms: *strata* and *stratification*.

_____

_____

_____

UNDERSTANDING KEY IDEAS

_____ **2.** Which of the following is an organic sedimentary rock?
  a. chemical limestone
  b. shale
  c. fossiliferous limestone
  d. conglomerate

3. Explain the process by which clastic sedimentary rock forms.

_____

_____

_____

_____

4. Describe the three main categories of sedimentary rock.

_____

_____

_____

_____

MATH SKILLS

5. A layer of a sedimentary rock is 2 m thick. How many years did it take for this layer to form if an average of 4 mm of sediment accumulated per year? Show your work below.

CRITICAL THINKING

6. Identifying Relationships Rocks are classified based on texture and composition. Which of these two properties would be more important for classifying clastic sedimentary rock?

_____

_____

_____

_____

7. Analyzing Processes Why do you think raindrop impressions are more likely to be preserved in fine-grained sedimentary rock rather than coarse-grained sedimentary rock?

_____

_____

_____

_____

Section Review

Metamorphic Rock
USING KEY TERMS

1. In your own words, define the following terms: *foliated* and *nonfoliated*.

_____

_____

_____

_____

UNDERSTANDING KEY IDEAS

_____ **2.** Which of the following is not a type of foliated metamorphic rock?
   a. gneiss
   b. slate
   c. marble
   d. schist

3. Explain the difference between contact metamorphism and regional metamorphism.

_____

_____

_____

_____

4. Explain how index minerals allow a scientist to understand the history of a metamorphic rock.

_____

_____

_____

_____

MATH SKILLS

5. For every 3.3 km a rock is buried, the pressure placed upon it increases 0.1 gigapascal (100 million pascals). If rock undergoing metamorphosis is buried at 16 km, what is the pressure placed on that rock? (Hint: The pressure of Earth's surface is .101 gigapascal.) Show your work below.

CRITICAL THINKING

6. Making Inferences If you had two metamorphic rocks, one that has garnet crystals and the other that has chlorite crystals, which one could have formed at a deeper level in the Earth's crust? Explain your answer.

_____

_____

_____

7. Applying Concepts Which do you think would be easier to break, a foliated rock, such as slate, or a nonfoliated rock, such as quartzite? Explain.

_____

_____

_____

8. Analyzing Processes A mountain range is located at a boundary where two tectonic plates are colliding. Would most of the metamorphic rock in the mountain range be a product of contact metamorphism or regional metamorphism? Explain.

_____

_____

_____

_____

Chapter Review

USING KEY TERMS

1. In your own words, write a definition for the term *rock cycle*.

_____

_____

Complete each of the following sentences by choosing the correct term from the word bank.

stratification   foliated
extrusive igneous rock texture

2. The _____ of a rock is determined by the sizes, shapes, and positions of the minerals the rock contains.

3. _____ metamorphic rock contains minerals that are arranged in plates or bands.

4. The most characteristic property of sedimentary rock

is _____.

5. _____ forms plains called *lava plateaus*.

UNDERSTANDING KEY IDEAS

Multiple Choice

_____ **6.** Sedimentary rock is classified into all of the following main categories except
   a. clastic sedimentary rock.
   b. chemical sedimentary rock.
   c. nonfoliated sedimentary rock.
   d. organic sedimentary rock.

_____ **7.** An igneous rock that cools very slowly has a _____ texture.
   a. foliated
   b. fine-grained
   c. nonfoliated
   d. coarse-grained

Chapter Review *continued*

_____ **8.** Igneous rock forms when
   a. minerals crystallize from a solution.
   b. sand grains are cemented together.
   c. magma cools and solidifies.
   d. mineral grains in a rock recrystallize.

_____ **9.** A _____ is a common structure found in metamorphic rock.
   a. ripple mark   **c.** sill
   b. fold   **d.** layer

_____ **10.** The process in which sediment is removed from its source and transported is called
   a. deposition.   **c.** weathering.
   b. erosion.   **d.** uplift.

_____ **11.** Mafic rocks are
   a. light-colored rocks rich in calcium, iron, and magnesium.
   b. dark-colored rocks rich in aluminum, potassium, silica, and sodium.
   c. light-colored rocks rich in aluminum, potassium, silica, and sodium.
   d. dark-colored rocks rich in calcium, iron, and magnesium.

Short Answer

12. Explain how composition and texture are used by scientists to classify rocks.

_____

_____

13. Describe two ways a rock can undergo metamorphism.

_____

_____

14. Explain why some minerals only occur in metamorphic rocks.

_____

_____

15. Describe how each type of rock changes as it moves through the rock cycle.

_____

_____

16. Describe two ways rocks were used by early humans and ancient civilizations.

_____

_____

CRITICAL THINKING

17. Concept Mapping Use the following terms to construct a concept map: *rocks, metamorphic, sedimentary, igneous, foliated, nonfoliated, organic, clastic, chemical, intrusive,* and *extrusive*.

18. Making Inferences If you were looking for fossils in the rocks around your home and the rock type that was closest to your home was metamorphic, do you think that you would find many fossils? Explain your answer.

_____

_____

_____

_____

_____

19. Applying Concepts Imagine that you want to quarry, or mine, granite. You have all of the equipment, but you have two pieces of land to choose from. One area has a granite batholith underneath it. The other has a granite sill. If both intrusive bodies are at the same depth, which one would be the better choice for you to quarry? Explain your answer.

_____

_____

_____

_____

_____

20. Applying Concepts The sedimentary rock coquina is made up of pieces of seashells. Which of the three kinds of sedimentary rock could coquina be? Explain your answer.

_____

_____

_____

_____

21. Analyzing Processes If a rock is buried deep inside the Earth, which geological processes cannot change the rock? Explain your answer.

_____

_____

_____

_____

| Chapter Review *continued*

INTERPRETING GRAPHICS

The bar graph below shows the percentage of minerals by mass that compose a sample of granite. Use the graph below to answer the questions that follow.

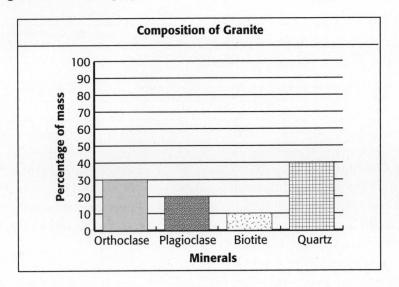

22. Your rock sample is made of four minerals. What percentage of each mineral makes up your sample?

_____

23. Both plagioclase and orthoclase are feldspar minerals. What percentage of the minerals in your sample of granite are not feldspar minerals?

_____

24. If your rock sample has a mass of 10 g, how many grams of quartz does it contain?

_____

25. Use paper, a compass, and a protractor or a computer to make a pie chart. Show the percentage of each of the four minerals your sample of granite contains. (Look in the Appendix of the student edition for help on making a pie chart.)

_____

_____

_____

_____

_____

Skills Worksheet

Section Review

Natural Resources

USING KEY TERMS

1. Use each of the following terms in a separate sentence: *natural resource, renewable resource, nonrenewable resource,* and *recycling.*

_____

_____

_____

_____

_____

UNDERSTANDING KEY IDEAS

2. How do humans use most natural resources?

_____

_____

_____

_____

_____ **3.** Which of the following is a renewable resource?

    a. oil

    b. water

    c. coal

    d. natural gas

4. Describe three ways to conserve natural resources.

_____

_____

_____

_____

_____

MATH SKILLS

5. If a faucet dripped for 8.6 h and 3.3 L of water dripped out every hour, how many liters of water dripped out altogether? Show your work below.

CRITICAL THINKING

6. Making Inferences How does human activity affect Earth's renewable and nonrenewable resources?

_____

_____

_____

_____

7. Applying Concepts List five products you regularly use that can be recycled.

_____

_____

_____

_____

8. Making Inferences Why is the availability of some renewable resources more of a concern now than it was 100 years ago?

_____

_____

_____

_____

Skills Worksheet

Section Review

Fossil Fuels

USING KEY TERMS

1. Use each of the following terms in a separate sentence: *energy resource, fossil fuel, petroleum, natural gas, coal, acid precipitation,* and *smog.*

_____

_____

_____

_____

_____

_____

UNDERSTANDING KEY IDEAS

_____ **2.** Which of the following types of coal contains the highest carbon content?
  a. lignite
  b. anthracite
  c. peat
  d. bituminous coal

3. Name a solid fossil fuel, a liquid fossil fuel, and a gaseous fossil fuel.

_____

_____

_____

4. Briefly describe how petroleum and natural gas form.

_____

_____

_____

_____

5. How do we obtain petroleum and natural gas?

_____

6. Describe the advantages and disadvantages of fossil fuel use.

_____

_____

_____

CRITICAL THINKING

7. Making Comparisons What is the difference between the organic material from which coal forms and the organic material from which petroleum and natural gas form?

_____

_____

_____

_____

8. Making Inferences Why can't carpooling and using mass-transit systems eliminate the problems associated with fossil fuels?

_____

_____

_____

_____

Name _____ Class _____ Date _____

INTERPRETING GRAPHICS

Use the pie chart below to answer the questions that follow.

Oil Production by Region

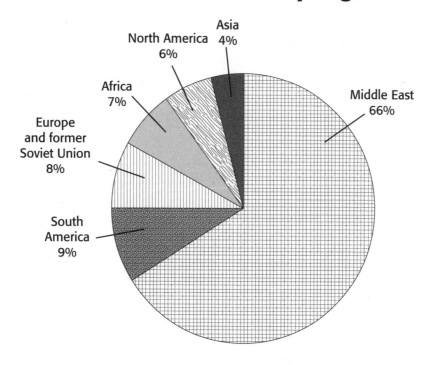

Source: International Energy Agency

9. Which region produces the most oil?

_____

_____

10. If the total sales of oil in 2002 were $500 billion, what was the value of the oil produced in North America?

_____

_____

_____

_____

Skills Worksheet)

Section Review

Alternative Resources

USING KEY TERMS

1. In your own words, write a definition for each of the following terms: *nuclear energy, solar energy, wind power, hydroelectric energy, biomass, gasohol,* and *geothermal energy.*

_____

_____

_____

_____

_____

_____

_____

UNDERSTANDING KEY IDEAS

_____ 2. Which of the following alternative resources requires hydrogen and oxygen to produce energy?
   a. fuel cells
   b. solar energy
   c. nuclear energy
   d. geothermal energy

3. Describe two ways of using solar energy.

_____

_____

4. Where is the production of hydroelectric energy practical?

_____

_____

5. Describe two ways to release biomass energy.

_____

_____

6. Describe two ways to use geothermal energy.

_____

_____

CRITICAL THINKING

7. Analyzing Methods If you were going to build a nuclear power plant, why wouldn't you build it in the middle of a desert?

_____

_____

_____

_____

8. Predicting Consequences If an alternative resource could successfully replace crude oil, how might the use of that resource affect the environment?

_____

_____

_____

_____

INTERPRETING GRAPHICS

Use the graph below to answer the questions that follow.

How Energy Is Used In the United States

Residential
19%

Industrial
38%

Commercial
16%

Transportattion
27%

9. What is the total percentage of energy that is used for commercial and industrial purposes?

_____

_____

10. What is the total percentage of energy that is not used for residential purposes?

_____

_____

Skills Worksheet

Chapter Review

USING KEY TERMS

The statements below are false. For each statement, replace the underlined term to make a true statement.

1. A liquid mixture of complex hydrocarbon compounds is called <u>natural gas</u>.

   _____

2. Energy that is released when a chemical compound reacts to produce a new compound is called <u>nuclear energy</u>.

   _____

   _____

For each pair of terms, explain how the meanings of the terms differ.

3. *solar energy* and *wind power*

   _____

   _____

4. *biomass* and *gasohol*

   _____

   _____

UNDERSTANDING KEY IDEAS

Multiple Choice

_____ 5. Which of the following resources is a renewable resource?
   a. coal
   b. trees
   c. oil
   d. natural gas

_____ 6. Which of the following fuels is NOT made from petroleum?
   a. jet fuel
   b. lignite
   c. kerosene
   d. fuel oil

_____ 7. Peat, lignite, and anthracite are all forms of
   a. petroleum.
   b. natural gas.
   c. coal.
   d. gasohol.

_____ **8.** Which of the following factors contributes to smog?
  a. automobiles
  b. sunlight
  c. mountains surrounding urban areas
  d. All of the above

_____ **9.** Which of the following resources is produced by fusion?
  a. solar energy   **c.** nuclear energy
  b. natural gas   **d.** petroleum

_____ **10.** To produce energy, nuclear power plants use a process called
  a. fission.   **c.** fractionation.
  b. fusion.   **d.** None of the above

_____ **11.** A solar-powered calculator uses
  a. solar collectors.   **c.** solar mirrors.
  b. solar panels.   **d.** solar cells.

Short Answer

12. How does acid precipitation form?

_____

_____

_____

13. If sunlight is free, why is electrical energy from solar cells expensive?

_____

_____

_____

14. Describe three ways that humans use natural resources.

_____

_____

_____

15. Explain how fossil fuels are found and obtained.

_____

_____

_____

| Chapter Review *continued*

CRITICAL THINKING

16. Concept Mapping Use the following terms to create a concept map:
fossil fuels, wind energy, energy resources, biomass, renewable resources, solar energy, nonrenewable resources, natural gas, gasohol, coal, and *oil.*

17. Predicting Consequences How would your life be different if fossil fuels were less widely available?

_____

_____

18. Evaluating Assumptions Are fossil fuels nonrenewable? Explain.

_____

_____

19. Evaluating Assumptions Why do we need to conserve renewable resources even though they can be replaced?

_____

_____

20. Evaluating Data What might limit the productivity of a geothermal power plant?

_____

_____

21. Identifying Relationships Explain why the energy we get from many of our resources ultimately comes from the sun.

_____

_____

22. Applying Concepts Describe the different ways you can conserve natural resources at home.

_____

_____

23. Identifying Relationships Explain why coal usually forms in different locations from where petroleum and natural gas form.

_____

_____

24. Applying Concepts Choose an alternative energy resource that you think should be developed more. Explain the reason for your choice.

_____

_____

INTERPRETING GRAPHICS

Use the graph below to answer the questions that follow.

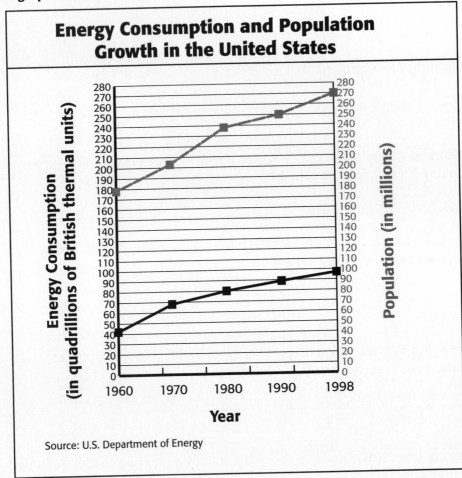

25. How many British thermal units were consumed in 1970?

_____

26. In what year was the most energy consumed?

_____

27. Why do you think that energy consumption has not increased at the same rate as the population has increased?

_____

_____

_____

_____

_____

Skills Worksheet

Section Review

Earth's Story and Those Who First Listened

USING KEY TERMS

1. Use each of the following terms in a separate sentence: *uniformitarianism*, *catastrophism*, and *paleontology*.

_____

_____

_____

UNDERSTANDING KEY IDEAS

_____ 2. Which of the following words describes change according to the principle of uniformitarianism?
   a. sudden
   b. rare
   c. global
   d. gradual

3. What is the difference between uniformitarianism and catastrophism?

_____

_____

4. Describe how the science of geology has changed.

_____

_____

5. Give one example of catastrophic global change.

_____

_____

6. Describe the work of three types of paleontologists.

_____

_____

_____

MATH SKILLS

7. An impact crater left by an asteroid strike has a radius of 85 km. What is the area of the crater? (Hint: The area of a circle is πr^2.) Show your work below.

CRITICAL THINKING

8. Analyzing Ideas Why is uniformitarianism considered the foundation behind modern geology?

_____

_____

_____

_____

_____

9. Applying Concepts Give an example of a type of recent catastrophe.

_____

_____

_____

_____

_____

Skills Worksheet

Section Review

Relative Dating: Which Came First?

USING KEY TERMS

1. In your own words, write a definition for each of the following terms: *relative dating*, *superposition*, and *geologic column*.

_____

_____

_____

_____

_____

UNDERSTANDING KEY IDEAS

_____ **2.** Molten rock that squeezes into existing rock and cools is called a(n)
   a. fold.
   b. fault.
   c. intrusion.
   d. unconformity.

3. List two events and two features that can disturb rock-layer sequences.

_____

_____

_____

_____

4. Explain how physical features are used to determine relative ages.

_____

_____

_____

_____

CRITICAL THINKING

5. Analyzing Concepts Is there a place on Earth that has all the layers of the geologic column? Explain.

_____

_____

_____

_____

6. Analyzing Methods Disconformities are hard to recognize because all the layers are horizontal. How does a geologist know when he or she is looking at a disconformity?

_____

_____

_____

_____

INTERPRETING GRAPHICS

7. Look at the diagram below. If the top rock layer were eroded and deposition later resumed, what type of unconformity would mark the boundary between older rock layers and the newly deposited rock layers?

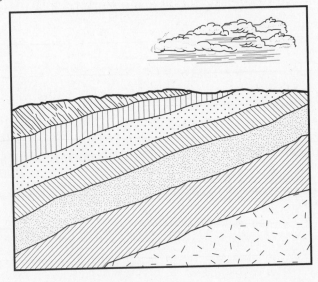

_____

_____

Section Review

Absolute Dating: A Measure of Time

USING KEY TERMS

1. Use each of the following terms in a separate sentence: *absolute dating*, *isotopes*, and *half-life*.

_____

_____

_____

_____

_____

UNDERSTANDING KEY IDEAS

_____ 2. Rubidium-87 has a half-life of
   a. 5,730 years.
   b. 4.5 billion years.
   c. 49 billion years.
   d. 1.3 billion years.

3. Explain how radioactive decay occurs.

_____

_____

_____

_____

4. How does radioactive decay relate to radiometric dating?

_____

_____

_____

_____

Section Review *continued*

5. List four types of radiometric dating.

_____

_____

_____

MATH SKILLS

6. A radioactive isotope has a half-life of 1.3 billion years. After 3.9 billion years, how much of the parent material will be left? Show your work below.

CRITICAL THINKING

7. Analyzing Methods Explain why radioactive decay must be constant in order for radiometric dating to be accurate.

_____

_____

_____

8. Applying Concepts Which radiometric-dating method would be most appropriate for dating artifacts found at Effigy Mounds? Explain.

_____

_____

_____

_____

Section Review

Looking at Fossils

USING KEY TERMS

Complete each of the following sentences by choosing the correct term from the word bank.

cast   index fossils   mold
trace fossils

1. A _____ is a cavity in rock where a plant or animal was buried.

2. _____ can be used to establish the age of rock layers.

UNDERSTANDING KEY IDEAS

_____ 3. Fossils are most often preserved in
     a. ice.
     b. amber.
     c. asphalt.
     d. rock.

4. Describe three types of trace fossils.

_____

_____

5. Explain how an index fossil can be used to date rock.

_____

_____

_____

6. Explain why the fossil record contains an incomplete record of the history of life on Earth.

_____

_____

_____

_____

| **Section Review** *continued*

7. Explain how fossils can be used to determine the history of changes in environments and organisms.

_____

_____

_____

MATH SKILLS

8. If a scientist finds the remains of a plant between a rock layer that contains 400 million-year-old *Phacops* fossils and a layer that contains 230 million-year-old *Tropites* fossils, how old could the plant fossil be? Show your work below.

CRITICAL THINKING

9. Making Inferences If you find rock layers containing fish fossils in a desert, what can you infer about the history of the desert?

_____

_____

_____

_____

10. Identifying Bias Because information in the fossil record is incomplete, scientists are left with certain biases concerning fossil preservation. Explain two of these biases.

_____

_____

_____

_____

_____

Skills Worksheet

Section Review

Time Marches On

USING KEY TERMS

1. Use each of the following terms in the same sentence: *era, period,* and *epoch.*

_____

_____

_____

_____

_____

_____

UNDERSTANDING KEY IDEAS

_____ **2.** The unit of geologic time that began 65.5 million years ago and
continues to the present is called the
   a. Holocene epoch.
   b. Cenozoic era.
   c. Phanerozoic eon.
   d. Quaternary period.

3. What are the major time intervals represented by the geologic time scale?

_____

4. Explain how geologic time is recorded in rock layers.

_____

_____

_____

_____

5. What kinds of environmental changes cause mass extinctions?

_____

_____

_____

_____

CRITICAL THINKING

6. Making Inferences What future event might mark the end of the Cenozoic era?

_____

_____

_____

7. Identifying Relationships How might a decrease in competition between species lead to the sudden appearance of many new species?

_____

_____

_____

_____

INTERPRETING GRAPHICS

8. Look at the illustration below. On the Earth-history clock shown, 1 h equals 383 million years, and 1 min equals 6.4 million years. In millions of years, how much more time is represented by the Proterozoic eon than the Phanerozoic eon?

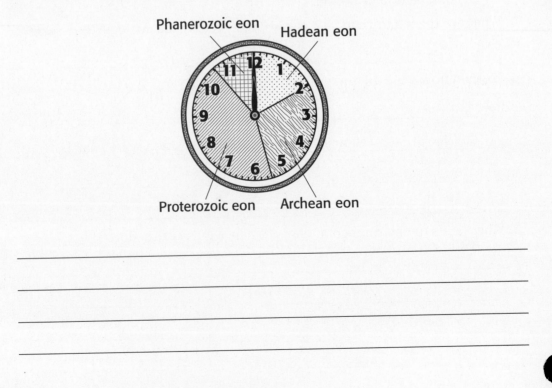

Phanerozoic eon   Hadean eon

Proterozoic eon   Archean eon

_____

_____

_____

Skills Worksheet

Chapter Review

USING KEY TERMS

1. In your own words, write a definition for each of the following terms: *superposition*, *geologic column*, and *geologic time scale*.

_____

_____

_____

For each pair of terms, explain how the meanings of the terms differ.

2. *uniformitarianism* and *catastrophism*

_____

_____

3. *relative dating* and *absolute dating*

_____

_____

4. *trace fossil* and *index fossil*

_____

_____

UNDERSTANDING KEY IDEAS

Multiple Choice

_____ **5.** Which of the following does not describe catastrophic change?
    a. widespread
    b. sudden
    c. rare
    d. gradual

_____ **6.** Scientists assign relative ages by using
    a. absolute dating.
    b. the principle of superposition.
    c. radioactive half-lives.
    d. carbon-14 dating.

Chapter Review *continued*

_____ **7.** Which of the following is a trace fossil?
  a. an insect preserved in amber
  b. a mammoth frozen in ice
  c. wood replaced by minerals
  d. a dinosaur trackway

_____ **8.** The largest divisions of geologic time are called
  a. periods.   **c.** eons.
  b. eras.   **d.** epochs.

_____ **9.** Rock layers cut by a fault formed
  a. after the fault.
  b. before the fault.
  c. at the same time as the fault.
  d. There is not enough information to determine the answer.

_____ **10.** Of the following isotopes, which is stable?
  a. uranium-238   **c.** carbon-12
  b. potassium-40   **d.** carbon-14

_____ **11.** A surface that represents a missing part of the geologic column is called a(n)
  a. intrusion.   **c.** unconformity.
  b. fault.   **d.** fold.

_____ **12.** Which method of radiometric dating is used mainly to date the remains of organisms that lived within the last 50,000 years?
  a. carbon-14 dating
  b. potassium-argon dating
  c. uranium-lead dating
  d. rubidium-strontium dating

Short Answer

13. Describe three processes by which fossils form.

_____

_____

_____

14. Identify the role of uniformitarianism in Earth science.

_____

_____

15. Explain how radioactive decay occurs.

_____

_____

16. Describe two ways in which scientists use fossils to determine environmental change.

_____

_____

17. Explain the role of paleontology in the study of Earth's history.

_____

_____

Chapter Review *continued*

CRITICAL THINKING

18. Concept Mapping Use the following terms to create a concept map: *age, half-life, absolute dating, radioactive decay, radiometric dating, relative dating, superposition, geologic column,* and *isotopes.*

19. Applying Concepts Identify how changes in environmental conditions can affect the survival of a species. Give two examples.

_____

_____

20. Identifying Relationships Why do paleontologists know more about hard-bodied organisms than about soft-bodied organisms?

_____

_____

21. Analyzing Processes Why isn't a 100 million-year-old fossilized tree made of wood?

_____

_____

INTERPRETING GRAPHICS

Use the diagram below to answer the questions that follow.

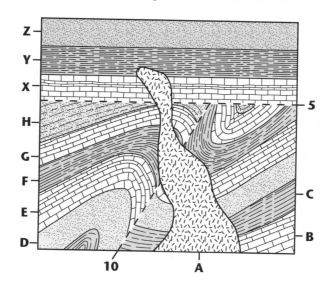

22. Is intrusion A younger or older than layer X? Explain.

_____

_____

23. What feature is marked by 5?

_____

_____

24. Is intrusion A younger or older than fault 10? Explain.

_____

_____

25. Other than the intrusion and faulting, what event happened in layers B, C, D, E, F, G, and H? Number this event, the intrusion, and the faulting in the order that they happened.

_____

_____

Skills Worksheet

Section Review

Inside the Earth
USING KEY TERMS
For each pair of terms, explain how the meanings of the terms differ.

1. *crust* and *mantle*

_____

_____

2. *lithosphere* and *asthenosphere*

_____

_____

UNDERSTANDING KEY IDEAS

_____ 3. The part of the Earth that is molten is the
   a. crust.
   b. mantle.
   c. outer core.
   d. inner core.

_____ 4. The part of the Earth on which the tectonic plates move is the
   a. lithosphere.
   b. asthenosphere.
   c. mesosphere.
   d. crust.

5. Identify the layers of the Earth by their chemical composition.

_____

_____

6. Identify the layers of the Earth by their physical properties.

_____

_____

7. Describe a tectonic plate.

_____

_____

| Section Review *continued*

8. Explain how scientists know about the structure of the Earth's interior.

_____

_____

INTERPRETING GRAPHICS

9. According to the wave speeds shown in the table below, which two physical layers of the Earth are densest?

Speed of Seismic Waves in Earth's Interior	
Physical layer	**Wave speed**
Lithosphere	7 to 8 km/s
Asthenosphere	7 to 11 km/s
Mesosphere	11 to 13 km/s
Outer core	8 to 10 km/s
Inner core	11 to 12 km/s

_____

_____

CRITICAL THINKING

10. Making Comparisons Explain the difference between the crust and the lithosphere.

_____

_____

_____

_____

11. Analyzing Ideas Why does a seismic wave travel faster through solid rock than through water?

_____

_____

_____

_____

Skills Worksheet

Section Review

Restless Continents

USING KEY TERMS

1. In your own words, write a definition for each of the following terms: *continental drift* and *sea-floor spreading*.

_____

_____

UNDERSTANDING KEY IDEAS

_____ 2. At mid-ocean ridges,
   a. the crust is older.
   b. sea-floor spreading occurs.
   c. oceanic lithosphere is destroyed.
   d. tectonic plates are colliding.

3. Explain how oceanic lithosphere forms at mid-ocean ridges.

_____

_____

4. What is magnetic reversal?

_____

_____

MATH SKILLS

5. If a piece of sea floor has moved 50 km in 5 million years, what is the yearly rate of sea-floor motion? Show your work below.

CRITICAL THINKING

6. Identifying Relationships Explain how magnetic reversals provide evidence for sea-floor spreading.

_____

_____

_____

_____

7. Applying Concepts Why do bands indicating magnetic reversals appear to be of similar width on both sides of a mid-ocean ridge?

_____

_____

_____

_____

8. Applying Concepts Why do you think that old rocks are rare on the ocean floor?

_____

_____

_____

_____

Skills Worksheet

Section Review

The Theory of Plate Tectonics

USING KEY TERMS

1. In your own words, write a definition for the term *plate tectonics*.

_____

_____

UNDERSTANDING KEY IDEAS

_____ **2.** The speed a tectonic plate moves per year is best measured in
  a. kilometers per year.
  b. centimeters per year.
  c. meters per year.
  d. millimeters per year.

3. Briefly describe three possible driving forces of tectonic plate movement.

_____

_____

_____

4. Explain how scientists use GPS to measure the rate of tectonic plate movement.

_____

_____

MATH SKILLS

5. If an orbiting satellite has a diameter of 60 cm, what is the total surface area of the satellite? (Hint: *surface area* $= 4\pi r^2$) Show your work below.

Section Review *continued*

CRITICAL THINKING

6. Identifying Relationships When convection takes place in the mantle, why does cool rock material sink and warm rock material rise?

_____

_____

_____

_____

7. Analyzing Processes Why does oceanic crust sink beneath continental crust at convergent boundaries?

_____

_____

_____

_____

Skills Worksheet

Section Review

Deforming the Earth's Crust

USING KEY TERMS

For each pair of key terms, explain how the meanings of the terms differ.

1. *compression* and *tension*

_____

_____

2. *uplift* and *subsidence*

_____

_____

UNDERSTANDING KEY IDEAS

_____ **3.** The type of fault in which the hanging wall moves up relative to the footwall is called a

  a. strike-slip fault.

  b. fault-block fault.

  c. normal fault.

  d. reverse fault.

4. Describe three types of folds.

_____

_____

_____

5. Describe three types of faults.

_____

_____

_____

6. Identify the most common types of mountains.

_____

_____

_____

7. What is rebound?

_____

_____

8. What are rift zones, and how do they form?

_____

_____

CRITICAL THINKING

9. Predicting Consequences If a fault occurs in an area where rock layers have been folded, which type of fault is it likely to be? Why?

_____

_____

_____

_____

10. Identifying Relationships Would you expect to see a folded mountain range at a mid-ocean ridge? Explain your answer.

_____

_____

_____

_____

Section Review *continued*

INTERPRETING GRAPHICS

Use the diagram below to answer the questions that follow.

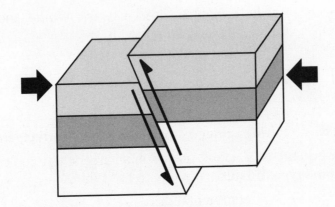

11. What type of fault is shown in the diagram?

_____

_____

12. At what kind of tectonic boundary would you most likely find this fault?

_____

_____

Skills Worksheet

Chapter Review

USING KEY TERMS

1. Use the following terms in the same sentence: *crust, mantle,* and *core.*

_____

_____

_____

Complete each of the following sentences by choosing the correct term from the word bank.

asthenosphere   uplift

tension   continental drift

2. The hypothesis that continents can drift apart and have done so in the past is

known as _____.

3. The _____ is the soft layer of the mantle on which the

tectonic plates move.

4. _____ is stress that occurs when forces act to stretch an

object.

5. The rising of regions of the Earth's crust to higher elevations is called

_____.

UNDERSTANDING KEY IDEAS

Multiple Choice

_____ **6.** The strong, lower part of the mantle is a physical layer called the
   a. lithosphere.
   b. mesosphere.
   c. asthenosphere.
   d. outer core.

_____ **7.** The type of tectonic plate boundary that forms from a collision
   between two tectonic plates is a
   a. divergent plate boundary.
   b. transform plate boundary.
   c. convergent plate boundary.
   d. normal plate boundary.

_____ **8.** The bending of rock layers due to stress in the Earth's crust is known as
   a. uplift.
   b. folding.
   c. faulting.
   d. subsidence.

_____ **9.** The type of fault in which the hanging wall moves up relative to the footwall is called a
   a. strike-slip fault.
   b. fault-block fault.
   c. normal fault.
   d. reverse fault.

_____ **10.** The type of mountain that forms when rock layers are squeezed together and pushed upward is the
   a. folded mountain.
   b. fault-block mountain.
   c. volcanic mountain.
   d. strike-slip mountain.

_____ **11.** Scientists' knowledge of the Earth's interior has come primarily from
   a. studying magnetic reversals in oceanic crust.
   b. using a system of satellites called the *global positioning system*.
   c. studying seismic waves generated by earthquakes.
   d. studying the pattern of fossils on different continents.

Short Answer

12. Explain how scientists use seismic waves to map the Earth's interior.

_____

_____

13. How do magnetic reversals provide evidence of sea-floor spreading?

_____

_____

14. Explain how sea-floor spreading provides a way for continents to move.

_____

_____

15. Describe two types of stress that deform rock.

_____

_____

16. What is the global positioning system (GPS), and how does GPS allow scientists to measure the rate of motion of tectonic plates?

_____

_____

CRITICAL THINKING

17. Concept Mapping Use the following terms to create a concept map: *sea-floor spreading, convergent boundary, divergent boundary, subduction zone, transform boundary,* and *tectonic plates.*

18. Applying Concepts Why does oceanic lithosphere sink at subduction zones but not at mid-ocean ridges?

_____

_____

_____

19. Identifying Relationships New tectonic material continually forms at divergent boundaries. Tectonic plate material is also continually destroyed in subduction zones at convergent boundaries. Do you think that the total amount of lithosphere formed on the Earth is about equal to the amount destroyed? Why?

_____

_____

_____

│ Chapter Review *continued*

20. **Applying Concepts** Folded mountains usually form at the edge of a tectonic plate. How can you explain folded mountain ranges located in the middle of a tectonic plate?

_____

_____

INTERPRETING GRAPHICS

Imagine that you could travel to the center of the Earth. Use the diagram below to answer the questions that follow.

Composition	Structure
Crust (50 km)	Lithosphere (150 km)
Mantle (2,900 km)	Asthenosphere (250 km)
	Mesosphere (2,550 km)
Core (3,430 km)	Outer core (2,200 km)
	Inner core (1,228 km)

21. How far beneath the Earth's surface would you have to go before you were no longer passing through rock that had the composition of granite?

_____

_____

22. How far beneath the Earth's surface would you have to go to find liquid material in the Earth's core?

_____

_____

23. At what depth would you find mantle material but still be within the lithosphere?

_____

_____

24. How far beneath the Earth's surface would you have to go to find solid iron and nickel in the Earth's core?

_____

_____

Skills Worksheet

Section Review

What Are Earthquakes?
USING KEY TERMS

Complete each of the following sentences by choosing the correct term from the word bank.

Deformation   P waves   Elastic rebound   S waves

1. _____ is the change in shape of rock due to stress.

2. _____ always travel ahead of other waves.

UNDERSTANDING KEY IDEAS

_____ **3.** Seismic waves that shear rock side to side are called
  a. surface waves.   **c.** P waves.
  b. P waves.   **d.** Both (b) and (c)

4. Where do earthquakes occur?

_____

_____

_____

5. What is the direct cause of earthquakes?

_____

_____

_____

6. Describe the three types of plate motion and the faults that are characteristic of each type of motion.

_____

_____

_____

7. What is an earthquake zone?

_____

_____

_____

| Section Review *continued*

MATH SKILLS

8. A seismic wave is traveling through the Earth at an average rate of speed of 8 km/s. How long will it take the wave to travel 480 km? Show your work below.

CRITICAL THINKING

9. Applying Concepts Given what you know about elastic rebound, why do you think some earthquakes are stronger than others?

_____

_____

_____

_____

10. Identifying Relationships Why are surface waves more destructive to buildings than P waves or S waves are?

_____

_____

_____

_____

11. Identifying Relationships Why do you think the majority of earthquake zones are located at tectonic plate boundaries?

_____

_____

_____

_____

Skills Worksheet

Section Review

Earthquake Measurement
USING KEY TERMS

1. In your own words, write a definition for each of the following terms: *epicenter* and *focus*.

_____

_____

UNDERSTANDING KEY IDEAS

2. What is the difference between a seismograph and a seismogram?

_____

_____

_____

3. Explain how earthquakes are detected.

_____

_____

_____

4. Briefly explain the steps of the S-P time method for locating an earthquake's epicenter.

_____

_____

_____

5. Why might an earthquake have more than one intensity value?

_____

_____

_____

| Section Review *continued*

MATH SKILLS

6. How much more ground motion is produced by an earthquake of magnitude 7.0 than by an earthquake of magnitude 4.0? Show your work below.

CRITICAL THINKING

7. Making Inferences Why is a 6.0 magnitude earthquake so much more destructive than a 5.0 magnitude earthquake?

_____

_____

_____

_____

8. Identifying Bias Which do you think is the more important measure of earthquakes, strength or intensity? Explain.

_____

_____

_____

_____

9. Making Inferences Do you think an earthquake of moderate magnitude can produce high Modified Mercalli intensity values?

_____

_____

_____

_____

Skills Worksheet

Section Review

Earthquakes and Society

USING KEY TERMS

1. In your own words, write a definition for each of the following terms: *gap hypothesis* and *seismic gap*.

_____

_____

UNDERSTANDING KEY IDEAS

_____ **2.** A weight that is placed on a building to make the building earthquake resistant is called a(n)

  a. active tendon system.   **c.** mass damper.
  b. cross brace.   **d.** base isolator.

3. How is an area's earthquake-hazard level determined?

_____

_____

4. Compare the strength and frequency method with the gap hypothesis method for predicting earthquakes.

_____

_____

_____

5. What is a common way of making homes more earthquake resistant?

_____

_____

_____

6. Describe four pieces of technology that are designed to make buildings earthquake resistant.

_____

_____

_____

_____

7. Name five items that you should store in case of an earthquake.

_____

_____

_____

_____

MATH SKILLS

8. Of the approximately 420,000 earthquakes recorded each year, about 140 have a magnitude greater than 6.0. What percentage of total earthquakes have a magnitude greater than 6.0? Show your work below.

CRITICAL THINKING

9. Evaluating Hypotheses Seismologists predict that there is a 20% chance that an earthquake of magnitude 7.0 or greater will fill a seismic gap during the next 50 years. Is the hypothesis incorrect if the earthquake does not happen? Explain your answer.

_____

_____

_____

10. Applying Concepts Why is a large earthquake often followed by numerous aftershocks?

_____

_____

_____

_____

Skills Worksheet

Chapter Review

USING KEY TERMS

1. Use each of the following terms in a separate sentence: *seismic wave,*
   P wave, and *S wave.*

   _____

   _____

For each pair of terms, explain how the meanings of the terms differ.

2. *seismograph* and *seismogram*

   _____

   _____

3. *epicenter* and *focus*

   _____

   _____

4. *gap hypothesis* and *seismic gap*

   _____

   _____

UNDERSTANDING KEY IDEAS
Multiple Choice

_____ 5. When rock is _____, energy builds up in it. Seismic waves occur as this
energy is _____.
   a. plastically deformed, increased
   b. elastically deformed, released
   c. plastically deformed, released
   d. elastically deformed, increased

_____ 6. Reverse faults are created
   a. by divergent plate motion.
   b. by convergent plate motion.
   c. by transform plate motion.
   d. All of the above

_____ 7. The last seismic waves to arrive are
   a. P waves.
   b. body waves.
   c. S waves.
   d. surface waves.

| Chapter Review *continued*

_____ **8.** If an earthquake begins while you are in a building, the safest thing for
you to do is
  a. to run out into an open space.
  b. to get under the strongest table, chair, or other piece of furniture.
  c. to call home.
  d. to crouch near a wall.

_____ **9.** How many major earthquakes (magnitude 7.0 to 7.9) happen on
average in the world each year?
  a. 1   **c.** 120
  b. 18   **d.** 800

_____ **10.** _____ counteract pressure that pushes and pulls at the side of a
building during an earthquake.
  a. Base isolators   **c.** Active tendon systems
  b. Mass dampers   **d.** Cross braces

Short Answer

11. Can the S-P time method be used with one seismograph station to locate the
epicenter of an earthquake? Explain your answer.

_____

_____

12. Explain how the Richter scale and the Modified Mercalli Intensity Scale are
different.

_____

_____

13. What is the relationship between the strength of earthquakes and earthquake
frequency?

_____

_____

14. Explain the way that different seismic waves affect rock as they travel
through it.

_____

_____

15. Describe some steps you can take to protect yourself and your property from
earthquakes.

_____

_____

CRITICAL THINKING

16. Concept Mapping Use the following terms to create a concept map: *focus, epicenter, earthquake start time, seismic waves, P waves,* and *S waves.*

17. Identifying Relationships Would a strong or light earthquake be more likely to happen along a major fault where there have not been many recent earthquakes? Explain. (Hint: Think about the average number of earthquakes of different magnitudes that occur annually.)

_____

_____

_____

_____

_____

18. Applying Concepts Japan is located near a point where three tectonic plates converge. What would you imagine the earthquake-hazard level in Japan to be? Explain why.

_____

_____

_____

19. Applying Concepts You learned that if you are in a car during an earthquake and are out in the open, it is best to stay in the car. Can you think of any situation in which you might want to leave a car during an earthquake?

_____

_____

_____

20. Identifying Relationships You use gelatin to simulate rock in an experiment in which you are investigating the way different seismic waves affect rock. In what ways is your gelatin model limited?

_____

_____

_____

_____

_____

| Chapter Review *continued*

INTERPRETING GRAPHICS

The graph below illustrates the relationship between earthquake magnitude and the height of tracings on a seismogram. Charles Richter initially formed his magnitude scale by comparing the heights of seismogram readings for different earthquakes. Use the graph below to answer the questions that follow.

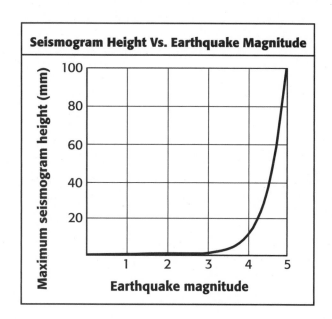

21. According to the graph, what would the magnitude of an earthquake be if its maximum seismogram height is 10 mm?

_____

22. According to the graph, what is the difference in maximum seismogram height (in mm) between an earthquake of magnitude 4.0 and an earthquake of magnitude 5.0?

_____

23. Look at the shape of the curve on the graph. What does this tell you about the relationship between seismogram heights and earthquake magnitudes? Explain.

_____

_____

_____

_____

Skills Worksheet

Section Review

Volcanic Eruptions

USING KEY TERMS

1. In your own words, write a definition for each of the following terms: *volcano*, *magma chamber*, and *vent*.

_____

_____

_____

UNDERSTANDING KEY IDEAS

_____ 2. Which of the following factors influences whether a volcano erupts explosively?
   a. the concentration of volcanic bombs in the magma
   b. the concentration of phosphorus in the magma
   c. the concentration of aa in the magma
   d. the concentration of water in the magma

3. How are lava and pyroclastic material classified? Describe four types of lava.

_____

_____

_____

_____

4. Which produces more pyroclastic material: an explosive eruption or a nonexplosive eruption?

_____

_____

5. Explain how the presence of silica and water in magma increases the chances of an explosive eruption.

_____

_____

6. What is a pyroclastic flow?

_____

_____

MATH SKILLS

7. A sample of magma is 64% silica. Express this percentage as a simplified fraction. Show your work below.

CRITICAL THINKING

8. Analyzing Ideas How is an explosive eruption similar to opening a can of soda that has been shaken? Be sure to describe the role of carbon dioxide.

_____

_____

_____

_____

9. Making Inferences Predict the silica content of aa, pillow lava, and blocky lava.

_____

_____

_____

_____

10. Making Inferences Explain why the names of many types of lava are Hawaiian but the names of many types of pyroclastic material are Italian and Indonesian.

_____

_____

_____

_____

_____

Name _____ Class _____ Date _____

Section Review

Effects of Volcanic Eruptions

USING KEY TERMS

Complete each of the following sentences by choosing the correct term from the word bank.

caldera   crater

1. A _____ is a funnel-shaped hole around the central vent.

2. A _____ results when a magma chamber partially empties.

UNDERSTANDING KEY IDEAS

_____ **3.** Which type of volcano results from alternating explosive and nonexplosive eruptions?

  a. composite volcano

  b. cinder cone volcano

  c. rift-zone volcano

  d. shield volcano

4. Why do cinder cone volcanoes have narrower bases and steeper sides than shield volcanoes do?

_____

_____

_____

_____

5. Why does a volcano's crater tend to get larger over time?

_____

_____

_____

_____

| Section Review *continued*

MATH SKILLS

6. The fastest lava flow recorded was 60 km/h. A horse can gallop as fast as 48 mi/h. Could a galloping horse outrun the fastest lava flow? (Hint: 1 km = 0.621 mi) Show your work below.

CRITICAL THINKING

7. Making Inferences Why did it take a year for the effects of the Tambora eruption to be experienced in New England?

_____

_____

_____

_____

Skills Worksheet

Section Review

Causes of Volcanic Eruptions

USING KEY TERMS

1. Use each of the following terms in a separate sentence: *hot spot* and *rift zone*.

_____

_____

UNDERSTANDING KEY IDEAS

_____ 2. If the temperature of a rock remains constant but the pressure on the
rock decreases, what tends to happen?
   a. The temperature increases.
   b. The rock becomes liquid.
   c. The rock becomes solid.
   d. The rock subducts.

_____ 3. Which of the following words is a synonym for *dormant*?
   a. predictable
   b. active
   c. dead
   d. sleeping

4. What is the Ring of Fire?

_____

_____

_____

5. Explain how convergent and divergent plate boundaries cause magma
formation.

_____

_____

_____

6. Describe four methods that scientists use to predict volcanic eruptions.

_____

_____

_____

| Section Review *continued*

7. Why does an oceanic plate tend to subduct when it collides with a continental plate?

_____

_____

_____

MATH SKILLS

8. If a tectonic plate moves at a rate of 2 km every 1 million years, how long would it take a hot spot to form a chain of volcanoes 100 km long? Show your work below.

CRITICAL THINKING

9. Making Inferences New crust is constantly being created at mid-ocean ridges. So, why is the oldest oceanic crust only about 150 million years old?

_____

_____

_____

_____

10. Identifying Relationships If you are studying a volcanic deposit, would the youngest layers be more likely to be found on the top or on the bottom? Explain your answer.

_____

_____

_____

_____

Skills Worksheet

Chapter Review

USING KEY TERMS

For each pair of terms, explain how the meanings of the terms differ.

1. *caldera* and *crater*

_____

_____

2. *lava* and *magma*

_____

_____

3. *lava* and *pyroclastic material*

_____

_____

4. *vent* and *rift*

_____

_____

5. *cinder cone volcano* and *shield volcano*

_____

_____

UNDERSTANDING KEY IDEAS

Multiple Choice

_____ **6.** The type of magma that tends to cause explosive eruptions has a
   a. high silica content and high viscosity.
   b. high silica content and low viscosity.
   c. low silica content and low viscosity.
   d. low silica content and high viscosity.

_____ **7.** Lava that flows slowly to form a glassy surface with rounded wrinkles is called
   a. aa lava.
   b. pahoehoe lava.
   c. pillow lava.
   d. blocky lava.

_____ **8.** Magma forms within the mantle most often as a result of
   a. high temperature and high pressure.
   b. high temperature and low pressure.
   c. low temperature and high pressure.
   d. low temperature and low pressure.

_____ **9.** What causes an increase in the number and intensity of small earthquakes before an eruption?
   a. the movement of magma
   b. the formation of pyroclastic material
   c. the hardening of magma
   d. the movement of tectonic plates

_____ **10.** If volcanic dust and ash remain in the atmosphere for months or years, what do you predict will happen?
   a. Solar reflection will decrease, and temperatures will increase.
   b. Solar reflection will increase, and temperatures will increase.
   c. Solar reflection will decrease, and temperatures will decrease.
   d. Solar reflection will increase, and temperatures will decrease.

_____ **11.** At divergent plate boundaries,
   a. heat from Earth's core causes mantle plumes.
   b. oceanic plates sink, which causes magma to form.
   c. tectonic plates move apart.
   d. hot spots cause volcanoes.

_____ **12.** A theory that helps explain the causes of both earthquakes and volcanoes is the theory of
   a. pyroclastics.   **c.** climactic fluctuation.
   b. plate tectonics.   **d.** mantle plumes.

Short Answer

13. How does the presence of water in magma affect a volcanic eruption?

_____

_____

_____

14. Describe four clues that scientists use to predict eruptions.

_____

_____

_____

_____

_____

Chapter Review *continued*

15. Identify the characteristics of the three types of volcanoes.

_____

_____

_____

16. Describe the positive effects of volcanic eruptions.

_____

_____

CRITICAL THINKING

17. Concept Mapping Use the following terms to create a concept map: *volcanic bombs, aa, pyroclastic material, pahoehoe, lapilli, lava,* and *volcano.*

18. Identifying Relationships You are exploring a volcano that has been dormant for some time. You begin to keep notes on the types of volcanic debris that you see as you walk. Your first notes describe volcanic ash. Later, your notes describe lapilli. In what direction are you most likely traveling—toward the crater or away from the crater? Explain your answer.

_____

_____

_____

_____

19. Making Inferences Loihi is a submarine Hawaiian volcano that might grow to form a new island. The Hawaiian Islands are located on the Pacific plate, which is moving northwest. Considering how this island chain may have formed, where do you think the new volcanic island will be located? Explain your answer.

_____

_____

_____

_____

20. Evaluating Hypotheses What evidence could confirm the existence of mantle plumes?

_____

_____

_____

_____

| Chapter Review *continued*

INTERPRETING GRAPHICS

The graph below illustrates the average change in temperature above or below normal for a community over several years. Use the graph below to answer the questions that follow.

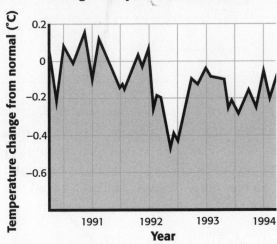

Average Temperature Variation

21. If the variation in temperature over the years was influenced by a major volcanic eruption, when did the eruption most likely take place? Explain.

_____

_____

_____

_____

22. If the temperature were measured only once each year (at the beginning of the year), how would your interpretation be different?

_____

_____

_____

_____

Skills Worksheet

Section Review

Weathering
USING KEY TERMS

1. In your own words, write a definition for each of the following terms: *weathering, mechanical weathering, abrasion, chemical weathering,* and *acid precipitation.*

_____

_____

_____

_____

_____

UNDERSTANDING KEY IDEAS

_____ **2.** Which of the following things cannot cause mechanical weathering?
   a. water
   b. acid
   c. wind
   d. animals

3. List three things that cause chemical weathering of rocks.

_____

_____

_____

4. Describe three ways abrasion occurs in nature.

_____

_____

_____

5. Describe the similarity in the ways tree roots and ice mechanically weather rock.

_____

_____

_____

6. Describe five sources of chemical weathering.

_____

_____

_____

_____

_____

CRITICAL THINKING

7. Making Inferences Why does acid precipitation weather rocks faster than normal precipitation?

_____

_____

_____

8. Making Comparisons Compare the weather processes that affect a rock on top of a mountain and a rock buried beneath the ground.

_____

_____

_____

MATH SKILLS

9. Substances that have a pH of less than 7 are acidic. For each pH unit lower, the acidity is ten times greater. For example, normal precipitation is slightly acidic at a 5.6 pH. If acid precipitation were measured at 4.6 pH, it would be 10 times more acidic than normal precipitation. How many times more acidic would precipitation at 3.6 pH be than normal precipitation? Show your work below.

Skills Worksheet

Section Review

Rates of Weathering

USING KEY TERMS

1. In your own words, write a definition for the term *differential weathering*.

_____

_____

UNDERSTANDING KEY IDEAS

_____ **2.** A rock will have a lower rate of weathering when the rock
   a. is in a humid climate.
   b. is a very hard rock, such as granite.
   c. is at a high elevation.
   d. has more surface area exposed to weathering.

3. How does surface area affect the rate of weathering?

_____

_____

4. How does climate affect the rate of weathering?

_____

_____

_____

5. Why does the peak of a mountain weather faster than the rocks at the bottom of the mountain?

_____

_____

_____

| Section Review *continued*

MATH SKILLS

6. The surface area of an entire cube is 96 cm^2. If the length and width of each side are equal, what is the length of one side of the cube? Show your work below.

CRITICAL THINKING

7. Making Inferences Does the rate of chemical weathering increase or stay the same when a rock becomes more mechanically weathered? Why?

_____

_____

_____

_____

Skills Worksheet

Section Review

From Bedrock to Soil

USING KEY TERMS

1. Use each of the following terms in a separate sentence: *soil, parent rock, bedrock, soil texture, soil structure, humus,* and *leaching.*

_____

_____

_____

_____

_____

_____

_____

UNDERSTANDING KEY IDEAS

_____ 2. Which of the following soil properties influences soil moisture?
   a. soil horizon
   b. soil fertility
   c. soil structure
   d. soil pH

_____ 3. Which of the following soil properties influences how nutrients can be dissolved in soil?
   a. soil texture
   b. soil fertility
   c. soil structure
   d. soil pH

4. When is parent rock the same as bedrock?

_____

_____

5. What is the difference between residual and transported soils?

_____

_____

Section Review *continued*

6. Which climate has the most thick, fertile soil?

_____

_____

7. How does soil temperature influence arctic soil?

_____

_____

MATH SKILLS

8. If a soil sample is 60% sand particles and has 30 million particles of soil, how many of those soil particles are sand? Show your work below.

CRITICAL THINKING

9. Identifying Relationships In which type of climate would leaching be more common—tropical rain forest or desert?

_____

10. Making Comparisons Although arctic climates are extremely different from desert climates, their soils may be somewhat similar. Explain why.

_____

_____

_____

Skills Worksheet

Section Review

Soil Conservation
USING KEY TERMS

1. In your own words, write a definition for each of the following terms: *soil conservation* and *erosion*.

_____

_____

_____

UNDERSTANDING KEY IDEAS

2. What are three important benefits that soil provides?

_____

_____

_____

_____ **3.** Practicing which of the following soil conservation techniques will replace nutrients in the soil?
   a. cover crop use   **c.** terracing
   b. no-till farming   **d.** contour plowing

4. How does crop rotation benefit soil?

_____

_____

_____

5. List five methods of soil conservation, and describe how each helps prevent the loss of soil.

_____

_____

_____

_____

_____

MATH SKILLS

6. Suppose it takes 500 years to form 2 cm of new soil without erosion. If a farmer needs at least 35 cm of soil to plant a particular crop, how many years will the farmer need to wait before planting his or her crop? Show your work below.

CRITICAL THINKING

7. Applying Concepts Why do land animals, even meat eaters, depend on soil to survive?

_____

_____

_____

_____

Chapter Review

USING KEY TERMS

1. In your own words, write a definition for each of the following terms: *abrasion* and *soil texture*.

_____

_____

2. Use each of the following terms in a separate sentence: *soil conservation* and *erosion*.

_____

_____

For each pair of terms, explain how the meanings of the terms differ.

3. *mechanical weathering* and *chemical weathering*

_____

_____

4. *soil* and *parent rock*

_____

_____

UNDERSTANDING KEY IDEAS

Multiple Choice

_____ **5.** Which of the following processes is a possible effect of water?
   a. mechanical weathering
   b. chemical weathering
   c. abrasion
   d. All of the above

_____ **6.** In which climate would you find the fastest rate of chemical weathering?
   a. a warm, humid climate
   b. a cold, humid climate
   c. a cold, dry climate
   d. a warm, dry climate

| Chapter Review *continued*

_____ **7.** Which of the following properties does soil texture affect?
  a. soil pH
  b. soil temperature
  c. soil consistency
  d. None of the above

_____ **8.** Which of the following properties describes a soil's ability to supply nutrients?
  a. soil structure
  b. infiltration
  c. soil fertility
  d. consistency

_____ **9.** Soil is important because it provides
  a. housing for animals.
  b. nutrients for plants.
  c. storage for water.
  d. All of the above

_____ **10.** Which of the following soil conservation techniques prevents erosion?
  a. contour plowing
  b. terracing
  c. no-till farming
  d. All of the above

Short Answer

11. Describe the two major types of weathering.

_____

_____

12. Why is Devil's Tower higher than the surrounding area?

_____

_____

13. Why is soil in temperate forests thick and fertile?

_____

_____

14. What can happen to soil when soil conservation is not practiced?

_____

_____

15. Describe the process of land degradation.

_____

_____

16. How do cover crops help prevent soil erosion?

_____

_____

CRITICAL THINKING

17. Concept Mapping Use the following terms to create a concept map: *weathering, chemical weathering, mechanical weathering, abrasion, ice wedging, oxidation,* and *soil.*

18. Analyzing Processes Heat generally speeds up chemical reactions. But weathering, including chemical weathering, is usually slowest in hot, dry climates. Why?

_____

_____

19. Making Inferences Mechanical weathering, such as ice wedging, increases surface area by breaking larger rocks into smaller rocks. Draw conclusions about how mechanical weathering can affect the rate of chemical weathering.

_____

_____

20. Evaluating Data A scientist has a new theory. She believes that climates that receive heavy rains all year long have thin topsoil. Given what you have learned, decide if the scientist's theory is correct. Explain your answer.

_____

_____

21. Analyzing Processes What forms of mechanical and chemical weathering would be most common in the desert? Explain your answer.

_____

_____

22. Applying Concepts If you had to plant a crop on a steep hill, what soil conservation techniques would you use to prevent erosion?

_____

_____

23. Making Comparisons Compare the weathering processes in a warm, humid climate with those in a dry, cold climate.

_____

_____

_____

| Chapter Review *continued*

INTERPRETING GRAPHICS

The graph below shows how the density of water changes when temperature changes. The denser a substance is, the less volume it occupies. In other words, as most substances get colder, they contract and become denser. But water is unlike most other substances. When water freezes, it expands and becomes less dense.

Use the graph below to answer the questions that follow.

The Density of Water

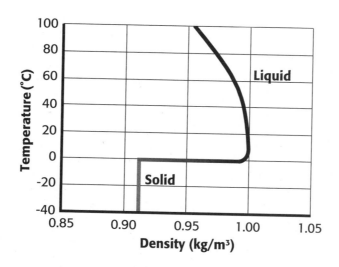

24. Which has the greater density: water at 40°C or water at –20°C?

_____

25. How would the line in the graph look if water behaved like most other liquids?

_____

_____

26. Which substance would be a more effective agent of mechanical weathering: water or another liquid? Why?

_____

_____

_____

_____

Skills Worksheet

Section Review

The Active River
USING KEY TERMS

1. Use each of the following terms in a separate sentence: *erosion, water cycle, tributary, watershed, divide, channel,* and *load.*

_____

_____

_____

_____

_____

_____

_____

_____

UNDERSTANDING KEY IDEAS

_____ **2.** Which of the following drains a watershed?
  a. a divide
  b. a drainage basin
  c. a tributary
  d. a water system

3. Describe how the Grand Canyon was formed.

_____

_____

4. Draw the water cycle. In your drawing, label *condensation, precipitation,* and *evaporation.*

Section Review *continued*

5. What are three factors that affect the rate of stream erosion?

_____

_____

_____

6. Which stage of river development is characterized by flat floodplains?

_____

_____

CRITICAL THINKING

7. Making Inferences How does the water cycle help develop river systems?

_____

_____

_____

_____

_____

_____

8. Making Comparisons How do youthful rivers, mature rivers, and old rivers differ?

_____

_____

_____

_____

_____

_____

| Section Review *continued*

INTERPRETING GRAPHICS

Use the pie graph shown in your textbook to answer the questions that follow.

9. Where is most of the water in the world found?

_____

_____

_____

10. In what form is the majority of the world's fresh water?

_____

_____

_____

Skills Worksheet

Section Review

Stream and River Deposits

USING KEY TERMS

1. In your own words, write a definition for each of the following terms: *deposition* and *floodplain*.

_____

_____

UNDERSTANDING KEY IDEAS

_____ 2. Which of the following forms at places in a river where the current slows?
   a. a placer deposit
   b. a delta
   c. a floodplain
   d. a levee

_____ 3. Which of the following can help to prevent a flood?
   a. a placer deposit
   b. a delta
   c. a floodplain
   d. a levee

4. Where do alluvial fans form?

_____

_____

5. Explain why floodplains are both good and bad areas for farming.

_____

_____

MATH SKILLS

6. A river flows at a speed of 8 km/h. If you floated on a raft in this river, how far would you have traveled after 5 h? Show your work below

Section Review *continued*

CRITICAL THINKING

7. Identifying Relationships What factors increase the likelihood that sediment will be deposited?

_____

_____

_____

_____

_____

8. Making Comparisons How are alluvial fans and deltas similar?

_____

_____

_____

_____

_____

_____

Skills Worksheet

Section Review

Water Underground
USING KEY TERMS

1. Use the following terms in the same sentence: *water table, aquifer, porosity,* and *artesian spring*.

_____

_____

_____

_____

UNDERSTANDING KEY IDEAS

_____ 2. Which of the following describes an aquifer's ability to allow water to flow through?
  a. porosity
  b. permeability
  c. geology
  d. recharge zone

3. What is a water table?

_____

_____

4. Describe how particles affect the porosity of an aquifer.

_____

_____

5. Explain the difference between an artesian spring and other springs.

_____

_____

6. Name a feature that is formed by underground erosion.

_____

_____

| **Section Review** *continued*

7. Name two features that are formed by underground deposition.

_____

_____

8. What type of weathering process causes underground erosion?

_____

_____

MATH SKILLS

9. Groundwater in an area flows at a speed of 4 km/h. How long would it take the water to flow 10 km to its spring? Show your work below.

CRITICAL THINKING

10. Predicting Consequences Explain how urban growth might affect the recharge zone of an aquifer.

_____

_____

_____

_____

11. Making Comparisons Explain the difference between a spring and a well.

_____

_____

_____

12. Analyze Relationships What is the relationship between the zone of aeration, the zone of saturation, and the water table?

_____

_____

_____

Skills Worksheet

Section Review

Using Water Wisely

USING KEY TERMS

1. Use each of the following terms in a separate sentence: *point-source pollution, nonpoint-source pollution, sewage treatment plant,* and *septic tank.*

_____

_____

_____

_____

UNDERSTANDING KEY IDEAS

_____ 2. Which of the following can help protect fish from acid rain?
   a. dissolved oxygen
   b. nitrates
   c. alkalinity
   d. point-source pollution

_____ 3. What type of wastewater treatment can be used for an individual home?
   a. sewage treatment plant
   b. primary treatment
   c. secondary treatment
   d. septic tank

4. Which kind of water pollution is often caused by runoff of fertilizers?

_____

_____

5. Describe what DO is.

_____

_____

6. What factors affect the level of dissolved oxygen in water?

_____

_____

7. Describe how water is conserved in industry.

_____

_____

MATH SKILLS

8. If 25% of water used in your home is used to water the lawn and you used a total of 95 gal of water today, how many gallons of water did you use to water the lawn? Show your work below.

CRITICAL THINKING

9. Making Inferences How do bacteria help break down the waste in water treatment plants?

_____

_____

_____

_____

10. Applying Concepts Other than examples listed in this section, what are some ways you can conserve water?

_____

_____

_____

_____

11. Making Inferences Why is it better to water your lawn at night instead of during the day?

_____

_____

_____

Chapter Review

USING KEY TERMS

The statements below are false. For each statement, replace the underlined term to make a true statement.

1. A stream that flows into a lake or into a larger stream is a <u>water cycle</u>.

_____

_____

2. The area along a river that forms from sediment deposited when the river overflows is a <u>delta</u>.

_____

_____

3. A rock's ability to let water pass through it is called <u>porosity</u>.

_____

_____

For each pair of terms, explain how the meanings of the terms differ.

4. *divide* and *watershed*

_____

_____

_____

5. *artesian springs* and *wells*

_____

_____

_____

6. *point-source pollution* and *nonpoint-source pollution*

_____

_____

_____

Chapter Review *continued*

UNDERSTANDING KEY IDEAS
Multiple Choice

_____ **7.** Which of the following processes is not part of the water cycle?
  a. evaporation
  b. percolation
  c. condensation
  d. deposition

_____ **8.** Which features are common in youthful river channels?
  a. meanders
  b. floodplains
  c. rapids
  d. sandbars

_____ **9.** Which depositional feature is found at the coast?
  a. delta
  b. floodplain
  c. alluvial fan
  d. placer deposit

_____ **10.** Caves are mainly a product of
  a. erosion by rivers.
  b. river deposition.
  c. water pollution.
  d. erosion by groundwater.

_____ **11.** Which of the following is necessary for aquatic life to survive?
  a. dissolved oxygen
  b. nitrate
  c. alkalinity
  d. point-source pollution

_____ **12.** During primary treatment at a sewage treatment plant,
  a. water is sent to an aeration tank.
  b. water is mixed with bacteria and oxygen.
  c. dirty water is passed through a large screen.
  d. water is sent to a settling tank where chlorine is added.

Short Answer

13. Identify and describe the location of the water table.

_____

_____

14. Explain how surface water enters an aquifer.

_____

_____

15. Why are caves usually found in limestone-rich regions?

_____

_____

_____

CRITICAL THINKING

16. Concept Mapping Use the following terms to create a concept map: *zone of aeration, zone of saturation, water table, gravity, porosity,* and *permeability.*

17. Identifying Relationships What is water's role in erosion and deposition?

_____

_____

_____

_____

18. Analyzing Processes What are the features of a river channel that has a steep gradient?

_____

_____

_____

_____

19. Analyzing Processes Why is groundwater hard to clean?

_____

_____

_____

_____

20. Evaluating Conclusions How can water be considered both a renewable and a nonrenewable resource? Give an example of each case.

_____

_____

_____

_____

21. Analyzing Processes Does water vapor lose or gain energy during the process of condensation? Explain.

_____

_____

_____

_____

INTERPRETING GRAPHICS

The hydrograph below illustrates data collected on river flow during field investigations over a period of 1 year. The discharge readings are from the Yakima River, in Washington. Use the hydrograph below to answer the questions that follow.

Hydrograph of the Yakima River

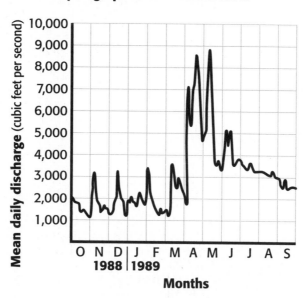

22. In which months is there the highest river discharge?

_____

_____

23. Why is there such a high river discharge during these months?

_____

_____

_____

_____

24. What might cause the peaks in river discharge between November and March?

_____

_____

_____

Skills Worksheet

Section Review

Shoreline Erosion and Deposition

USING KEY TERMS

Complete each of the following sentences by choosing the correct terms from the word bank.

shoreline   beach

1. A _____ is any area of the shoreline made up of material deposited by waves.

2. An area in which land and a body of water meet is

   a _____ .

UNDERSTANDING KEY IDEAS

_____ 3. Which of the following is a result of wave deposition?
   a. sea arch
   b. sea cave
   c. barrier spit
   d. headland

4. How do wave deposits affect a shoreline?

   _____

   _____

5. Describe how sand moves along a beach.

   _____

   _____

6. What are six shoreline features created by wave erosion?

   _____

   _____

   _____

   _____

   _____

   _____

Section Review *continued*

---

7. How can the energy of waves traveling through water affect a shoreline?

_____

_____

8. Would a small wave or a large wave have more energy? Explain your answer.

_____

_____

MATH SKILLS

9. Imagine that there is a large boulder on the edge of a shoreline. If the wave period is 15 s long, how many times is the boulder hit in a year? Show your work below.

CRITICAL THINKING

10. Applying Concepts Not all beaches are made from light-colored sand. Explain why this statement is true.

_____

_____

_____

11. Making Inferences How can severe storms over the ocean affect shoreline erosion and deposition?

_____

_____

_____

12. Making Predictions How could a headland change in 250 years? Describe some of the features that may form.

_____

_____

_____

Skills Worksheet

Section Review

Wind Erosion and Deposition

USING KEY TERMS

In each of the following sentences, replace the incorrect term with the correct term from the word bank.

dune   saltation
deflation   abrasion

1. Deflation hollows are mounds of wind-deposited sand.

_____

_____

2. The removal of fine sediment by wind is called abrasion.

_____

_____

UNDERSTANDING KEY IDEAS

_____ **3.** Which of the following landforms is the result of wind deposition?
   a. deflation hollow
   b. desert pavement
   c. dune
   d. abrasion

4. Describe how material is moved in areas where strong winds blow.

_____

_____

_____

_____

5. Explain the process of abrasion.

_____

_____

_____

_____

_____

MATH SKILLS

6. Data collected by a team of scientists studying dunes show that a particular dune moves about 40 m per year. How far does this dune move in 1 day? Show your work below

CRITICAL THINKING

7. Identifying Relationships Explain the relationship between plant cover and wind erosion.

_____

_____

_____

_____

8. Applying Concepts If you climbed up the steep side of a sand dune, is it likely that you traveled in the direction the wind was blowing? Explain.

_____

_____

_____

_____

Skills Worksheet)

Section Review

Erosion and Deposition by Ice
USING KEY TERMS

Complete each of the following sentences by choosing the correct term from the word bank.

glacial drift   glacier
stratified drift   till

1. A glacial deposit that is sorted into layers based on the size of the rock

material is called _____.

2. _____ is all of the material carried and deposited by

glaciers.

3. Unsorted rock material that is deposited directly by the ice when it melts is

_____.

4. A _____ is an enormous mass of moving ice.

UNDERSTANDING KEY IDEAS

_____ **5.** Which of the following is not a type of moraine?
     a. lateral
     b. horn
     c. ground
     d. medical

6. Explain the difference between alpine and continental glaciers.

_____

_____

7. Name five landscape features formed by alpine glaciers.

_____

_____

_____

_____

_____

| Section Review *continued*

8. Describe two ways in which glaciers move.

_____

_____

MATH SKILLS

9. A recent study shows that a glacier in Alaska is melting at a rate of 23 ft per year. At what rate is the glacier melting in meters? (Hint: 1 ft = 0.3 m)

CRITICAL THINKING

10. Analyzing Ideas Explain why continental glaciers smooth the landscape and alpine glaciers create a rugged landscape.

_____

_____

_____

_____

11. Applying Concepts How can a glacier deposit both sorted and unsorted material?

_____

_____

_____

_____

12. Applying Concepts Why are glaciers such effective agents of erosion and deposition?

_____

_____

_____

_____

Skills Worksheet

Section Review

The Effect of Gravity on Erosion and Deposition

USING KEY TERMS

Complete each of the following sentences by choosing the correct term from the word bank.

creep   mass movement
mudflow   landslide

1. A _____ occurs when a large amount of water mixes with

soil and rock.

2. The extremely slow movement of material downslope is

called _____.

UNDERSTANDING KEY IDEAS

_____ **3.** Which of the following is a factor that affects creep?
  a. water
  b. burrowing animals
  c. plant roots
  d. All of the above.

4. How is the angle of repose related to mass movement?

_____

_____

_____

_____

MATH SKILLS

5. If a lahar is traveling at 80 km/h, how long will it take the lahar to travel 20 km?

CRITICAL THINKING

6. Identifying Relationships Which types of mass movement are most dangerous to humans? Explain your answer.

_____

_____

_____

_____

7. Making Inferences How does deforestation increase the likelihood of mudflows?

_____

_____

_____

_____

Skills Worksheet

Chapter Review

USING KEY TERMS

For each pair of terms, explain how the meanings of the terms differ.

1. *shoreline* and *longshore current*

_____

_____

2. *beaches* and *dunes*

_____

_____

3. *deflation* and *saltation*

_____

_____

4. *continental glacier* and *alpine glacier*

_____

_____

5. *stratified drift* and *till*

_____

_____

6. *mudflow* and *creep*

_____

_____

UNDERSTANDING KEY IDEAS

Multiple Choice

_____ **7.** Surf refers to
   a. large storm waves in the open ocean.
   b. giant waves produced by hurricanes.
   c. breaking waves near the shoreline.
   d. small waves on a calm sea.

Chapter Review *continued*

_____ **8.** When waves cut completely through a headland, a _____ is formed.
    a. sea cave   **c.** wave-cut terrace
    b. sea cliff   **d.** sand bar

_____ **9.** A narrow strip of sand that is formed by wave deposition and is connected to the shore is called a
    a. barrier spit.   **c.** wave-cut terrace.
    b. sandbar.   **d.** headland.

_____ **10.** A wind-eroded depression is called a
    a. deflation hollow.   **c.** dune.
    b. desert pavement.   **d.** dust bowl.

_____ **11.** What term describes all types of glacial deposits?
    a. glacial drift   **c.** till
    b. dune   **d.** outwash

_____ **12.** Which of the following is NOT a landform created by an alpine glacier?
    a. cirque   **c.** horn
    b. deflation hollow   **d.** arête

_____ **13.** What is the term for a mass movement that is of volcanic origin?
    a. lahar   **c.** creep
    b. slump   **d.** rock fall

_____ **14.** Which of the following is a slow mass movement?
    a. mudflow   **c.** creep
    b. landslide   **d.** rock fall

Short Answer

15. Why do waves break when they near the shore?

_____

_____

16. Why are some areas more affected by wind erosion than other areas are?

_____

_____

17. What kind of mass movement happens continuously, day after day?

_____

_____

_____

| **Chapter Review** *continued*

18. In what direction do sand dunes move?

_____

19. Describe the different types of glacial moraines.

_____

_____

_____

_____

CRITICAL THINKING

20. Concept Mapping Use the following terms to create a concept map: *deflation, strong winds, saltation, dune,* and *desert pavement.*

21. Making Inferences How do humans increase the likelihood that wind erosion will occur?

_____

_____

_____

_____

22. Identifying Relationships If the large ice sheet covering Antarctica were to melt completely, what type of landscape would you expect Antarctica to have?

_____

_____

_____

_____

23. Applying Concepts You are a geologist who is studying rock to determine the direction of flow of an ancient glacier. What clues might help you determine the glacier's direction of flow?

_____

_____

_____

_____

24. Applying Concepts You are interested in purchasing a home that overlooks the ocean. The home that you want to buy sits atop a steep sea cliff. Given what you have learned about shoreline erosion, what factors would you take into consideration when deciding whether to buy the home?

_____

_____

_____

_____

Chapter Review *continued*

INTERPRETING GRAPHICS

The graph below illustrates coastal erosion and deposition at an imaginary beach over a period of 8 years. Use the graph below to answer the questions that follow.

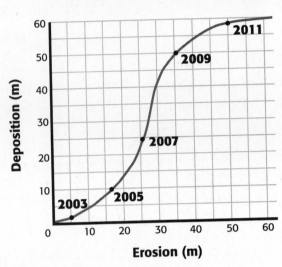

Erosion and Deposition (1998–2006)

25. What is happening to the beach over time?

_____

_____

26. In what year does the amount of erosion equal the amount of deposition?

_____

_____

27. Based on the erosion and deposition data for 2005, what might happen to the beach in the years that follow 2005?

_____

_____

_____

Section Review

Earth's Oceans

USING KEY TERMS

1. In your own words, write a definition for each of the following terms: *salinity* and *water cycle*.

_____

_____

UNDERSTANDING KEY IDEAS

_____ **2.** The top layer of ocean water that extends to 300 m below sea level is called the
    a. deep zone.
    b. surface zone.
    c. Gulf Stream.
    d. thermocline.

3. Name the major divisions of the global ocean.

_____

_____

4. Explain how Earth's first oceans formed.

_____

_____

5. Why is the ocean an important part of the water cycle?

_____

_____

6. Between which two steps of the water cycle does the ocean fit?

_____

_____

CRITICAL THINKING

7. Making Inferences Describe how the ocean plays a role in stabilizing Earth's weather conditions.

_____

_____

_____

| Section Review *continued*

8. **Identifying Relationships** List one factor that affects salinity in the ocean and one factor that affects ocean temperatures. Explain how each factor affects salinity or temperature.

_____

_____

_____

INTERPRETING GRAPHICS

Use the image in your textbook for this Section Review to answer the questions that follow.

9. At which stage would solid or liquid water fall to the Earth?

_____

10. At which stage would the sun's energy cause liquid to rise into the atmosphere as water vapor?

_____

Skills Worksheet

Section Review

The Ocean Floor

USING KEY TERMS

For each pair of terms, explain how the meanings of the terms differ.

1. *continental shelf* and *continental slope*

_____

_____

2. *abyssal plain* and *ocean trench*

_____

_____

3. *mid-ocean ridge* and *seamount*

_____

_____

UNDERSTANDING KEY IDEAS

_____ **4.** Sonar is a technology based on the
  a. *Geosat* satellite.
  b. surface currents in the ocean.
  c. zones of the ocean floor.
  d. echo-ranging behavior of bats.

5. List the two major regions of the ocean floor.

_____

6. Describe the subdivisions of the continental margin.

_____

_____

7. List three technologies for studying the ocean floor, and explain how they are used.

_____

_____

_____

_____

| Section Review *continued*

8. List three underwater missions that *Alvin* has been used for.

_____

_____

_____

9. Explain how *Jason II* and *Medea* are used to explore the ocean.

_____

_____

10. Describe how a bathymetric profile is made.

_____

_____

_____

MATH SKILLS

11. Air pressure at sea level is 1 atmosphere (atm). Under water, pressure increases by 1 atm every 10 m of depth. For example, at a depth of 10 m, water pressure is 2 atm. What is the pressure at 100 m? Show your work below.

CRITICAL THINKING

12. Making Comparisons How is exploring the oceans similar to exploring space?

_____

_____

_____

13. Applying Concepts Is the ocean floor a flat surface? Explain your answer.

_____

_____

_____

Skills Worksheet

Section Review

Life in the Ocean
USING KEY TERMS

The statements below are false. For each statement, replace the underlined term to make a true statement.

1. <u>Plankton</u> are organisms that swim actively in ocean water.

_____

2. The intertidal zone is part of the <u>pelagic zone</u>.

_____

3. Dolphins live in the <u>benthic environment</u>.

_____

UNDERSTANDING KEY IDEAS

_____ **4.** The deepest benthic zone is the
  a. pelagic environment.
  b. hadal zone.
  c. oceanic zone.
  d. abyssal zone.

5. List and briefly describe the three main groups of marine organisms.

_____

_____

_____

6. Name the two ocean environments. In your own words, describe where they are located in the ocean.

_____

_____

CRITICAL THINKING

7. Making Inferences Describe why organisms in the intertidal zone must be able to live underwater and on exposed land.

_____

_____

| Section Review *continued*

8. Applying Concepts How would the ocean's ecological zones change if sea level dropped 300 m?

_____

_____

INTERPRETING GRAPHICS

Use the diagram below to answer the following question.

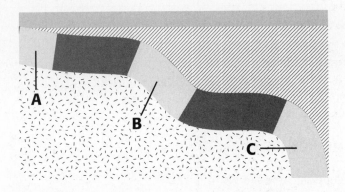

9. Identify the names of the ecological zones of the benthic environment shown above.

_____

_____

_____

Skills Worksheet

Section Review

Resources from the Ocean

USING KEY TERMS

1. In your own words, write a definition for the term *desalination*.

_____

UNDERSTANDING KEY IDEAS

_____ **2.** Mineral nodules on the ocean floor are
   a. renewable resources.
   b. easily mined.
   c. used during the process of desalination.
   d. nonliving resources.

3. List two ways of harvesting the ocean's living resources.

_____

_____

4. Name four nonliving resources in the ocean.

_____

_____

_____

_____

5. Explain how fish farms help meet the demand for fish.

_____

_____

_____

6. Explain how engineers decide where to drill for oil and natural gas in
   the ocean.

_____

_____

_____

| Section Review *continued*

MATH SKILLS

7. A kelp plant is 5 cm tall. If it grows at an average of 29 cm per day, how tall will the kelp plant be after 2 weeks? Show your work below.

CRITICAL THINKING

8. Analyzing Processes Explain why tidal energy and wave energy are considered renewable resources.

_____

_____

_____

9. Predicting Consequences Define the term *overfishing* in your own words. What would happen to the population of fish in the ocean if laws did not regulate overfishing? What would happen to the ocean ecosystem?

_____

_____

_____

10. Analyzing Ideas What is one benefit and one consequence of building a desalination plant? Would a desalination plant be beneficial to your local area? Explain why or why not.

_____

_____

_____

_____

_____

Skills Worksheet

Section Review

Ocean Pollution

USING KEY TERMS

1. Use the following terms in the same sentence: *point-source pollution* and *nonpoint-source pollution.*

_____

_____

UNDERSTANDING KEY IDEAS

_____ 2. Which of the following is an example of nonpoint-source pollution?
   a. a leak from an oil tanker
   b. a trash barge
   c. an unlined landfill
   d. water discharged by industries

3. List three types of ocean pollution. How can each of these types be prevented or minimized?

_____

_____

_____

4. Which part of raw sewage is a type of ocean pollution?

_____

_____

MATH SKILLS

5. Only 3% of Earth's water is drinkable. What portion of Earth's water is not drinkable? Show your work below.

6. A ship spilled 750,000 barrels of oil when it accidentally struck a reef. The oil company was able to recover 65% of the oil spilled. How many barrels of oil were not recovered? Show your work below.

CRITICAL THINKING

7. Identifying Relationships List and describe three measures that governments have taken to control ocean pollution.

_____

_____

_____

_____

8. Evaluating Conclusions What were two effects of the *Exxon Valdez* oil spill? Describe two ways in which oil spills can be prevented.

_____

_____

_____

9. Applying Concepts List two examples of nonpoint-source pollution that occur in your area. Explain why they are nonpoint-source pollution.

_____

_____

_____

_____

| Section Review *continued*

10. Predicting Consequences How can trash dumping and sludge dumping affect the food chains in the ocean?

_____

_____

_____

_____

Skills Worksheet

Chapter Review

USING KEY TERMS

Complete each of the following sentences by choosing the correct term from the word bank.

continental shelf	nonpoint-source pollution	benthic environment
abyssal plain	continental slope	point-source pollution
salinity	desalination	

1. The region of the ocean floor that is closest to the shoreline is the

   _____.

2. _____ is the process of removing salt from sea water.

3. _____ is a measure of the amount of dissolved salts in a
   liquid.

4. The _____ is the broad, flat part of the deep-ocean basin.

5. The region near the bottom of a pond, lake, or ocean is called the

   _____.

6. Pollution that comes from many sources rather than a single specific source

   is called _____.

UNDERSTANDING KEY IDEAS

Multiple Choice

_____ 7. The largest ocean is the
   a. Indian Ocean.
   b. Pacific Ocean.
   c. Atlantic Ocean.
   d. Arctic Ocean.

_____ 8. One of the most abundant elements in the ocean is
   a. potassium.
   b. calcium.
   c. chlorine.
   d. magnesium.

_____ 9. Which of the following affects the ocean's salinity?
   a. fresh water added by rivers
   b. currents
   c. evaporation
   d. All of the above

_____ 10. Most precipitation falls
   a. on land.
   b. into lakes and rivers.
   c. into the ocean.
   d. in rain forests.

_____ 11. Which of the following is a nonrenewable resource in the ocean?
   a. fish
   b. tidal energy
   c. oil
   d. All of the above

_____12. Which benthic zone has a depth range between 200 m and 4,000 m?
  a. the bathyal zone   **c.** the hadal zone
  b. the abyssal zone   **d.** the sublittoral zone

_____13. The ocean floor and all of the organisms that live on or in it is the
  a. benthic environment.   **c.** neritic zone.
  b. pelagic environment.   **d.** oceanic zone.

Short Answer

14. Why does coastal water in areas that have hotter, drier climates typically have a higher salinity than coastal water in cooler, more humid areas does?

_____

_____

_____

15. Describe two technologies used for studying the ocean floor.

_____

_____

16. Identify the two major regions of the ocean floor, and describe how the continental shelf, the continental slope, and the continental rise are related.

_____

_____

17. In your own words, write a definition for each of the following terms: *plankton, nekton,* and *benthos.* Give two examples of each.

_____

_____

_____

_____

18. List two living resources and two nonliving resources that are harvested from the ocean.

_____

_____

_____

CONCEPT MAPPING

19. Concept Mapping Use the following terms to create a concept map:
water cycle, evaporation, condensation, precipitation, atmosphere, and *oceans.*

Chapter Review *continued*

20. Making Inferences What benefit other than being able to obtain fresh water from salt water comes from desalination?

_____

_____

21. Making Comparisons Explain the difference between a bathymetric profile and a seismic reading.

_____

_____

22. Analyzing Ideas In your own words, define *nonpoint-source pollution* and *point-source pollution*. Give an example of each. What is being done to control ocean pollution?

_____

_____

_____

_____

_____

_____

INTERPRETING GRAPHICS

The graph below shows the ecological zones of the ocean. Use the graph below to answer the questions that follow.

Ecological Zones of the Ocean

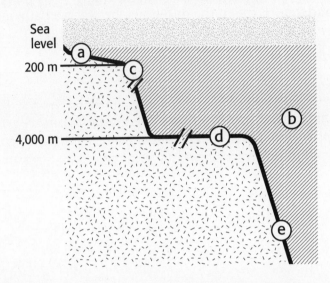

23. At which point would you most likely find an anglerfish?

_____

24. At which point would you most likely find tube worms?

_____

25. Which ecological zone is shown at point c?

_____

Which depth zone is shown at point c?

_____

26. Name an organism that you might find at point e.

_____

Section Review

Currents
USING KEY TERMS

The statements below are false. For each statement, replace the underlined word to make a true statement.

1. <u>Deep currents</u> are directly controlled by wind.

_____

2. An increase in density in parts of the ocean can cause <u>surface currents</u> to form.

_____

_____

UNDERSTANDING KEY IDEAS

_____ **3.** Surface currents
    a. are formed by wind.
    b. are streamlike movements of water.
    c. can travel across entire oceans.
    d. All of the above

4. List three factors that control surface currents.

_____

_____

_____

5. How does a continent affect the movement of a surface current?

_____

_____

_____

6. Explain how temperature and salinity affect the formation of deep currents.

_____

_____

_____

| Section Review *continued*

MATH SKILLS

7. The Gulf Stream flows along the North Carolina coast at 90 million cubic meters per second and at 40 million cubic meters per second when it turns eastward. How much faster is the Gulf Stream flowing along the coast than when it turns eastward? Show your work below.

CRITICAL THINKING

8. Evaluating Conclusions If there were no land on Earth's surface, what would the pattern of surface currents look like? Explain your answer.

_____

_____

_____

_____

9. Making Comparisons Compare the factors that contribute to the formation of surface currents and deep currents.

_____

_____

_____

_____

Skills Worksheet

Section Review

Currents and Climate

USING KEY TERMS

1. Use each of the following terms in a separate sentence: *upwelling*, *El Niño*, and *La Niña*.

_____

_____

_____

UNDERSTANDING KEY IDEAS

_____ 2. The Gulf Stream carries warm water to the North Atlantic Ocean, which contributes to
   a. a harsh winter in the British Isles.
   b. a cold-water surface current that flows to the British Isles.
   c. a mild climate for the British Isles.
   d. a warm-water surface current that flows along the coast of California.

3. Why might the climate in Scotland be relatively mild even though the country is located at a high latitude?

_____

_____

4. Name two disasters caused by El Niño.

_____

_____

MATH SKILLS

5. A fisher usually catches 540 kg of anchovies off the coast of Peru. During El Niño, the fisher caught 85% less fish. How many kilograms of fish did the fisher catch during El Niño? Show your work below.

CRITICAL THINKING

6. Applying Concepts Many marine organisms depend on upwelling to bring nutrients to the surface. How might El Niño affect a fisher's way of life?

_____

_____

_____

_____

Skills Worksheet

Section Review

Waves

USING KEY TERMS

For each pair of terms, explain how the meanings of the terms differ.

1. *whitecap* and *swell*

_____

_____

2. *undertow* and *longshore current*

_____

_____

3. *tsunami* and *storm surge*

_____

_____

UNDERSTANDING KEY IDEAS

_____ **4.** Longshore currents transport sediment
  a. to the open ocean.   **c.** only during low tide.
  b. along the shore.   **d.** only during high tide.

5. Where do deep-water waves become shallow-water waves?

_____

_____

6. Explain how water moves as waves travel through it.

_____

_____

_____

7. Name five events that can cause a tsunami.

_____

_____

_____

_____

| **Section Review** *continued*

8. Describe the two parts of a wave.

_____

_____

MATH SKILLS

9. If a barrier island that is 1 km wide and 10 km long loses 1.5 m of its width per year to erosion by a longshore current, how long will the island take to lose one-fourth of its width? Show your work below.

CRITICAL THINKING

10. Analyzing Processes How would you explain a bottle moving across the water in the same direction that the waves are traveling? Make a drawing of the bottle's movement.

_____

_____

_____

_____

Section Review *continued*

11. Analyzing Processes Describe the motion of a wave as it approaches the shore.

_____

_____

_____

_____

12. Applying Concepts Explain how energy plays a role in the creation of ocean waves.

_____

_____

_____

_____

13. Making Comparisons How does the formation of an undertow differ from the formation of a longshore current? How is sand on the beach affected by each?

_____

_____

_____

_____

Section Review

Tides

USING KEY TERMS

1. In your own words, write a definition for each of the following terms: *spring tides* and *neap tides*.

_____

_____

UNDERSTANDING KEY IDEAS

_____ **2.** Tides are at their highest during
   a. spring tide.
   b. neap tide.
   c. a tidal bore.
   d. the daytime.

3. Which tides have minimum tidal range? Which tides have maximum tidal range?

_____

_____

_____

4. What causes tidal ranges?

_____

_____

_____

MATH SKILLS

5. If it takes 24 h and 50 min for a spot on Earth that is facing the moon to rotate to face the moon again, how many minutes does it take? Show your work below.

CRITICAL THINKING

6. Applying Concepts How many days pass between the minimum and the maximum of the tidal range in any given area? Explain your answer.

_____

_____

_____

_____

7. Analyzing Processes Explain how the position of the moon relates to the occurrence of high tides and low tides.

_____

_____

_____

_____

Skills Worksheet

Chapter Review

USING KEY TERMS

For each pair of terms, explain how the meanings of the terms differ.

1. *surface current* and *deep current*

_____

_____

2. *El Niño* and *La Niña*

_____

_____

3. *spring tide* and *neap tide*

_____

_____

4. *tide* and *tidal range*

_____

_____

UNDERSTANDING KEY IDEAS

Multiple Choice

_____ **5.** Deep currents form when
   a. cold air decreases water density.
   b. warm air increases water density.
   c. the ocean surface freezes and solids from the water underneath are removed.
   d. salinity increases.

_____ **6.** When waves come near the shore,
   a. they speed up.   **c.** their wavelength increases.
   b. they maintain their speed.   **d.** their wave height increases.

_____ **7.** Whitecaps break
   a. in the surf.   **c.** in the open ocean.
   b. in the breaker zone.   **d.** as their wavelength increases.

_____ **8.** Tidal range is greatest during
   a. spring tide.   **c.** a tidal bore.
   b. neap tide.   **d.** the daytime.

_____ **9.** Tides alternate between high and low because the moon revolves around the Earth
  a. at the same speed the Earth rotates.
  b. at a much faster speed than the Earth rotates.
  c. at a much slower speed than the Earth rotates.
  d. at different speeds.

_____ **10.** El Niño can cause
  a. droughts to occur in Indonesia and Australia.
  b. upwelling to occur off the coast of South America.
  c. earthquakes.
  d. droughts to occur in the southern half of the United States.

Short Answer

11. Explain the relationship between upwelling and El Niño.

_____

_____

12. Describe the two parts of a wave. Describe how these two parts relate to wavelength and wave height.

_____

_____

13. Compare the relative positions of the Earth, moon, and sun during the spring and neap tides.

_____

_____

14. Explain the difference between the breaker zone and the surf.

_____

_____

15. Describe how warm-water currents affect the climate in the British Isles.

_____

_____

16. Describe the factors that form deep currents.

_____

_____

CRITICAL THINKING

17. Concept Mapping Use the following terms to create a concept map: *wind, deep currents, sun's gravity, types of ocean-water movement, surface currents, tides, increasing water density, waves,* and *moon's gravity.*

18. Identifying Relationships Why are tides more noticeable in Earth's oceans than on its land?

_____

_____

_____

_____

19. Expressing Opinions Explain why it's important to study El Niño and La Niña.

_____

_____

_____

_____

20. Applying Concepts Suppose you and a friend are planning a fishing trip to the ocean. Your friend tells you that the fish bite more in his secret fishing spot during low tide. If low tide occurred at the spot at 7 a.m. today and you are going to fish there in 1 week, at what time will low tide occur in that spot?

_____

_____

_____

21. Identifying Relationships Describe how global winds, the Coriolis effect, and continental deflections form a pattern of surface currents on Earth.

_____

_____

_____

_____

| Chapter Review *continued*

INTERPRETING GRAPHICS

The diagram below shows some of Earth's major surface currents that flow in the Western Hemisphere. Use the diagram to answer the questions that follow.

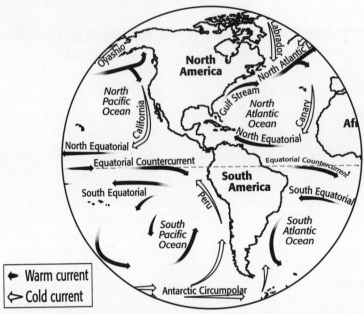

22. List two warm-water currents and two cold-water currents.

_____

_____

23. How do you think the Labrador Current affects the climate of Canada and Greenland?

_____

_____

_____

Skills Worksheet

Section Review

Characteristics of the Atmosphere
USING KEY TERMS

1. Use each of the following terms in a separate sentence: *air pressure,*
   atmosphere, troposphere, stratosphere, mesosphere, and *thermosphere.*

_____

_____

_____

_____

_____

_____

UNDERSTANDING KEY IDEAS

_____ 2. Why does the temperature of different layers of the atmosphere vary?
   a. because air temperature increases as altitude increases
   b. because the amount of energy radiated from the sun varies
   c. because of interference by humans
   d. because of the composition of gases in each layer

3. Why does air pressure decrease as altitude increases?

_____

_____

4. How can the thermosphere have high temperatures but not feel hot?

_____

_____

5. What determines the temperature of atmospheric layers?

_____

_____

6. What two gases make up most of the atmosphere?

_____

_____

Section Review *continued*

MATH SKILLS

7. If an average cloud has a density of 0.5 g/m^3 and has a volume of 1,000,000,000 m^3, what is the weight of an average cloud? Show your work below.

CRITICAL THINKING

8. Applying Concepts Apply what you know about about the relationship between altitude and air pressure to explain why rescue helicopters have a difficult time flying at altitudes over 6,000 m.

_____

_____

_____

_____

9. Making Inferences If the upper atmosphere is very thin, why do space vehicles heat up as they enter the atmosphere?

_____

_____

_____

_____

10. Making Inferences Explain why gases such as helium can escape Earth's atmosphere.

_____

_____

_____

_____

Skills Worksheet

Section Review

Atmospheric Heating
USING KEY TERMS

1. Use each of the following terms in a separate sentence: *thermal conduction, radiation, convection, greenhouse effect,* and *global warming.*

_____

_____

_____

_____

UNDERSTANDING KEY IDEAS

_____ 2. Which of the following is the best example of conduction?
   a. a light bulb warming a lampshade
   b. an egg cooking in a frying pan
   c. water boiling in a pot
   d. gases circulating in the atmosphere

3. Describe three ways that energy is transferred in the atmosphere.

_____

_____

_____

4. What is the difference between the greenhouse effect and global warming?

_____

_____

5. What is the radiation balance?

_____

_____

MATH SKILLS

6. Find the average of the following temperatures: 73.2°F, 71.1°F, 54.6°F, 65.5°F, 78.2°F, 81.9°F, and 82.1°F. Show your work below.

CRITICAL THINKING

7. Identifying Relationships How does the process of convection rely on radiation?

_____

_____

_____

_____

8. Applying Concepts Describe global warming in terms of the radiation balance.

_____

_____

_____

_____

Section Review

Global Winds and Local Winds
USING KEY TERMS

1. In your own words, write a definition for each of the following terms: *wind, Coriolis effect, jet stream, polar easterlies, westerlies,* and *trade winds.*

_____

_____

_____

_____

_____

_____

_____

UNDERSTANDING KEY IDEAS

_____ 2. Why does warm air rise and cold air sink?
   a. because warm air is less dense than cold air
   b. because warm air is denser than cold air
   c. because cold air is less dense than warm air
   d. because warm air has less pressure than cold air does

3. What are pressure belts?

_____

_____

4. What causes winds?

_____

_____

5. How does the Coriolis effect affect wind movement?

_____

_____

6. How are sea and land breezes similar to mountain and valley breezes?

_____

_____

| **Section Review** *continued*

7. Would there be winds if the Earth's surface were the same temperature every-where? Explain your answer.

_____

_____

MATH SKILLS

8. Flying an airplane at 500 km/h, an airplane pilot plans to reach her destination in 5 h. But she finds a jet stream moving 250 km/h in the direction she is traveling. If she gets a boost from the jet stream for 2 h, how long will the flight last? Show your work below.

CRITICAL THINKING

9. Making Inferences In the Northern Hemisphere, why do westerlies flow from the west but trade winds flow from the east?

_____

_____

_____

10. Applying Concepts Imagine you are near an ocean in the daytime. You want to go to the ocean, but you don't know how to get there. How might a local wind help you find the ocean?

_____

_____

_____

_____

Skills Worksheet

Section Review

Air Pollution

USING KEY TERMS

The statements below are false. For each statement, replace the underlined term to make a true statement.

1. <u>Air pollution</u> is a sudden change in the acidity of a stream or lake.

_____

_____

2. <u>Smog</u> is rain, sleet, or snow that has a high concentration of acid.

_____

_____

UNDERSTANDING KEY IDEAS

_____ 3. Which of the following statements describes the formation of smog?
   a. Acids in the air react with ozone.
   b. Ozone reacts with vehicle exhaust.
   c. Vehicle exhaust reacts with sunlight and ozone.
   d. Water vapor reacts with sunlight and ozone.

4. What is the difference between primary and secondary pollutants?

_____

_____

_____

_____

5. Describe five sources of indoor air pollution. Is all indoor air pollution caused by humans? Explain.

_____

_____

6. What is the ozone hole, and why does it form?

_____

_____

7. Describe five effects of air pollution on human health. How can air pollution be reduced?

_____

_____

_____

_____

_____

CRITICAL THINKING

8. Expressing Opinions How do you think that nations should resolve air-pollution problems that cross national boundaries?

_____

_____

_____

9. Making Inferences Why might establishing a direct link between air pollution and health problems be difficult?

_____

_____

_____

_____

INTERPRETING GRAPHICS

The map below shows the pH of precipitation measured at field stations in the northeastern U.S. On the pH scale, lower numbers indicate solutions that are more acidic than solutions with higher numbers. Use the map to answer the questions below.

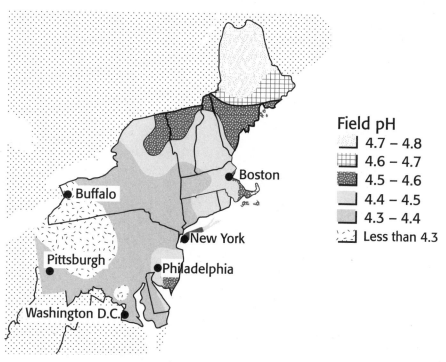

10. Which areas have the most acid precipitation? Hypothesize why.

_____

_____

_____

_____

11. Boston is a larger city than Buffalo is, but the precipitation in Buffalo is more acidic than the precipitation in Boston. Explain why.

_____

_____

_____

_____

_____

Chapter Review

USING KEY TERMS

For each pair of terms, explain how the meanings of the terms differ.

1. *air pressure* and *wind*

_____

2. *troposphere* and *thermosphere*

_____

3. *greenhouse effect* and *global warming*

_____

4. *convection* and *thermal conduction*

_____

5. *global wind* and *local wind*

_____

6. *stratosphere* and *mesosphere*

_____

UNDERSTANDING KEY IDEAS

Multiple Choice

_____ **7.** What is the most abundant gas in the atmosphere?
    a. oxygen   **c.** nitrogen
    b. hydrogen   **d.** carbon dioxide

_____ **8.** A major source of oxygen for the Earth's atmosphere is
    a. sea water.   **c.** plants.
    b. the sun.   **d.** animals.

_____ **9.** The bottom layer of the atmosphere, where almost all weather occurs, is the
    a. stratosphere.   **c.** thermosphere.
    b. troposphere.   **d.** mesosphere.

_____ **10.** What percentage of the solar energy that reaches the outer atmosphere is absorbed at the Earth's surface?
    a. 20%   **c.** 50%
    b. 30%   **d.** 70%

_____11. The ozone layer is located in the
  a. stratosphere.
  b. troposphere.
  c. thermosphere.
  d. mesosphere.

_____12. By which method does most thermal energy in the atmosphere circulate?
  a. conduction
  b. convection
  c. advection
  d. radiation

_____13. The balance between incoming and outgoing energy is called
  a. the convection balance.
  b. the conduction balance.
  c. the greenhouse effect.
  d. the radiation balance.

_____14. In which wind belt is most of the United States located?
  a. westerlies
  b. northeast trade winds
  c. southeast trade winds
  d. doldrums

_____15. Which of the following pollutants is NOT a primary pollutant?
  a. car exhaust
  b. acid precipitation
  c. smoke from a factory
  d. fumes from burning plastic

_____16. The Clean Air Act
  a. controls the amount of air pollutants that can be released from many sources.
  b. requires cars to run on fuels other than gasoline.
  c. requires many industries to use scrubbers.
  d. Both (a) and (c)

Short Answer

17. Why does the atmosphere become less dense as altitude increases?

_____

_____

_____

18. Explain why air rises when it is heated.

_____

_____

_____

19. What is the main cause of temperature changes in the atmosphere?

_____

_____

| Chapter Review *continued*

20. What are secondary pollutants, and how do they form? Give an example of a secondary pollutant.

_____

_____

_____

CRITICAL THINKING

21. Concept Mapping Use the following terms to create a concept map:
mesosphere, stratosphere, layers, temperature, troposphere, atmosphere.

22. Identifying Relationships What is the relationship between the greenhouse effect and global warming?

_____

_____

_____

_____

_____

_____

_____

23. Applying Concepts How do you think the Coriolis effect would change if the Earth rotated twice as fast as it does? Explain.

_____

_____

_____

_____

_____

_____

_____

24. Making Inferences The atmosphere of Venus has a very high level of carbon dioxide. How might this fact influence the greenhouse effect on Venus?

_____

_____

_____

_____

_____

_____

_____

| **Chapter Review** *continued*

INTERPRETING GRAPHICS

Use the diagram below to answer the questions that follow. When answering the questions that follow, assume that ocean currents do not affect the path of the boats.

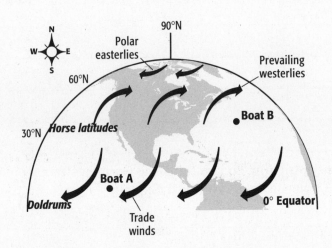

25. If Boat A traveled to 50°N, from which direction would the prevailing wind blow?

_____

26. If Boat B sailed with the prevailing westerlies in the Northern Hemisphere, in which direction would the boat be traveling?

_____

Skills Worksheet

Section Review

Water in the Air
USING KEY TERMS

1. In your own words, write a definition for each of the following terms: *relative humidity, condensation, cloud,* and *precipitation.*

_____

_____

_____

_____

UNDERSTANDING KEY IDEAS

_____ 2. Which of the following clouds is most likely to produce light to heavy, continuous rain?
   a. cumulus cloud
   b. cumulonimbus cloud
   c. nimbostratus cloud
   d. cirrus cloud

3. How is relative humidity affected by the amount of water vapor in the air?

_____

_____

4. What does a relative humidity of 75% mean?

_____

_____

5. Describe the path of water through the water cycle.

_____

_____

6. What are four types of precipitation?

_____

_____

_____

_____

CRITICAL THINKING

7. Applying Concepts Why are some clouds formed from water droplets, while others are made up of ice crystals?

_____

_____

8. Applying Concepts How can rain and hail fall from the same cumulonimbus cloud?

_____

_____

9. Identifying Relationships What happens to relative humidity as the air temperature drops below the dew point?

_____

_____

INTERPRETING GRAPHICS

Use the image in your textbook on the Section Review page to answer the following questions.

10. What type of cloud is shown in the image?

_____

11. How is this type of cloud formed?

_____

_____

12. What type of weather can you expect when you see this type of cloud? Explain.

_____

_____

Skills Worksheet)
Section Review

Air Masses and Fronts
USING KEY TERMS
For each pair of terms, explain how the meanings of the terms differ.

1. *front* and *air mass*

_____

_____

2. *cyclone* and *anticyclone*

_____

_____

UNDERSTANDING KEY IDEAS

_____ 3. What kind of front forms when a cold air mass displaces a warm air mass?

    a. a cold front   **c.** an occluded front

    b. a warm front   **d.** a stationary front

4. What are the major air masses that influence the weather in the United States?

_____

_____

5. What is one source region of a maritime polar air mass?

_____

_____

6. What are the characteristics of an air mass whose two-letter symbol is cP?

_____

_____

7. What are the four major types of fronts?

_____

_____

_____

_____

|Section Review *continued*

8. How do fronts cause weather changes?

_____

_____

9. How do cyclones and anticyclones affect the weather?

_____

_____

_____

MATH SKILLS

10. A cold front is moving toward the town of La Porte at 35 km/h. The front is 200 km away from La Porte. How long will it take for the front to get to La Porte? Show your work below.

CRITICAL THINKING

11. Applying Concepts How do air masses that form over the land and ocean affect weather in the United States?

_____

_____

_____

12. Identifying Relationships Why does the Pacific Coast have cool, wet winters and warm, dry summers? Explain.

_____

_____

_____

13. Applying Concepts Which air masses influence the weather where you live? Explain.

_____

_____

_____

Skills Worksheet

Section Review

Severe Weather

USING KEY TERMS

Complete each of the following sentences by choosing the correct term from the word bank.

hurricane	storm surge
tornado	lightning

1. Thunderstorms are very active electrically and often

cause _____.

2. A _____ forms when a funnel cloud pokes through

the bottom of a cumulonimbus cloud and makes contact with the ground.

UNDERSTANDING KEY IDEAS

_____ 3. The safest thing to do if you are caught outdoors during a tornado is to
   a. stay near buildings and roads.
   b. head for an open area.
   c. seek shelter near a large tree.
   d. None of the above

4. Describe how tornadoes form.

_____

_____

_____

_____

5. At what latitudes do hurricanes usually form?

_____

_____

6. What is lightning? What happens when lightning strikes?

_____

_____

_____

_____

▌Section Review *continued*

CRITICAL THINKING

7. Applying Concepts What items do you think you would need in a disaster kit? Explain.

_____

_____

8. Identifying Relationships What happens to a hurricane as it moves over land? Explain.

_____

_____

INTERPRETING GRAPHICS

Use the diagram in your textbook for this Section Review to answer the following questions.

9. Describe what is happening at point C.

_____

_____

10. What is point B?

_____

_____

11. What kind of weather can you expect at point A?

_____

_____

Section Review

Forecasting the Weather

USING KEY TERMS

1. In your own words, write a definition for each of the following terms: *thermometer*, *barometer*, and *anemometer*.

_____

_____

_____

_____

_____

_____

_____

_____

UNDERSTANDING KEY IDEAS

_____ 2. Which of the following instruments measures air pressure?
   a. thermometer
   b. barometer
   c. anemometer
   d. windsock

3. How does radar help meteorologists forecast the weather?

_____

_____

_____

_____

4. What does a station model represent?

_____

_____

_____

_____

MATH SKILLS

5. If it is 75°F outside, what is the temperature in degrees Celsius?
(Hint: °F = (°C × 9/ 5) + 32) Show your work below.

CRITICAL THINKING

6. Applying Concepts Why would a meteorologist compare a new weather map
with one that is 24 h old?

_____

_____

_____

_____

7. Making Inferences In the United States, why is weather data gathered from a
large number of station models?

_____

_____

_____

_____

8. Making Inferences How might several station models from different regions
plotted on a map help a meteorologist?

_____

_____

_____

_____

_____

Skills Worksheet

Chapter Review

USING KEY TERMS

For each pair of terms, explain how the meanings of the terms differ.

 1. *relative humidity* and *dew point*

_____

_____

 2. *condensation* and *precipitation*

_____

_____

 3. *air mass* and *front*

_____

_____

 4. *lightning* and *thunder*

_____

_____

 5. *tornado* and *hurricane*

_____

_____

 6. *barometer* and *anemometer*

_____

_____

UNDERSTANDING KEY IDEAS

Multiple Choice

 7. The process in which water changes from a liquid to gas is called
    a. precipitation.   **c.** evaporation.
    b. condensation.   **d.** water vapor.

 8. What is the relative humidity of air at its dew point?
    a. 0%   **c.** 75%
    b. 50%   **d.** 100%

9. Which of the following is NOT a type of condensation?

    a. fog

    b. cloud

    c. snow

    d. dew

10. High clouds made of ice crystals are called _____ clouds.

    a. stratus

    b. cumulus

    c. nimbostratus

    d. cirrus

11. Large thunderhead clouds that produce precipitation are called _____ clouds.

    a. nimbostratus

    b. cumulonimbus

    c. cumulus

    d. stratus

12. Strong updrafts within a thunderhead can produce

    a. snow.

    b. rain.

    c. sleet.

    d. hail.

13. A maritime tropical air mass contains

    a. warm, wet air.

    b. cold, moist air.

    c. warm, dry air.

    d. cold, dry air.

14. A front that forms when a warm air mass is trapped between cold air masses and is forced to rise is a(n)

    a. stationary front.

    b. warm front.

    c. occluded front.

    d. cold front.

15. A severe storm that forms as a rapidly rotating funnel cloud is called a

    a. hurricane.

    b. tornado.

    c. typhoon.

    d. thunderstorm.

16. The lines connecting points of equal air pressure on a weather map are called

    a. contour lines.

    b. highs.

    c. isobars.

    d. lows.

Short Answer

17. Explain the relationship between condensation and dew point.

_____

_____

18. Describe the conditions along a stationary front.

_____

_____

19. What are the characteristics of an air mass that forms over the Gulf of Mexico?

_____

_____

_____

20. Explain how a hurricane forms.

_____

_____

21. Describe the water cycle, and explain how it affects weather.

_____

_____

22. List the major similarities and differences between hurricanes and tornadoes.

_____

_____

23. Explain how a tornado forms.

_____

_____

24. Describe an interaction between weather and ocean systems.

_____

_____

25. What is a station model? What types of information do station models provide?

_____

_____

26. What type of technology is used to locate and measure the amount of precipitation in an area?

_____

_____

27. List two ways to keep yourself informed during severe weather.

_____

_____

28. Explain why staying away from floodwater is important even when the water is shallow.

_____

_____

CRITICAL THINKING

29. Concept Mapping Use the following terms to create a concept map: *evaporation, relative humidity, water vapor, dew, psychrometer, clouds,* and *fog.*

30. Making Inferences If both the air temperature and the amount of water vapor in the air change, is it possible for the relative humidity to stay the same? Explain.

_____

_____

_____

_____

31. Applying Concepts What can you assume about the amount of water vapor in the air if there is no difference between the wet- and dry-bulb readings of a psychrometer?

_____

_____

_____

_____

Name _____ Class _____ Date _____

32. Identifying Relationships Explain why the concept of relative humidity is important to understanding weather.

_____

_____

_____

INTERPRETING GRAPHICS

Use the weather map below to answer the questions that follow.

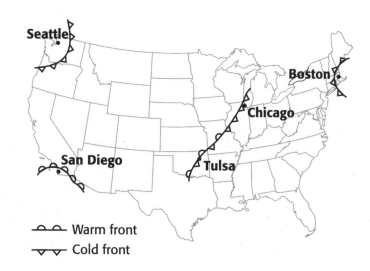

—⌒⌒— Warm front

—▽▽— Cold front

33. Where are thunderstorms most likely to occur? Explain your answer.

_____

_____

_____

_____

34. What are the weather conditions in Tulsa, Oklahoma? Explain your answer.

_____

_____

_____

_____

_____

Name _____ Class _____ Date _____

Skills Worksheet

Section Review

What Is Climate?

USING KEY TERMS

1. In your own words, write a definition for each of the following terms: *weather, climate, latitude, prevailing winds, elevation, surface current,* and *biome.*

_____

_____

_____

_____

_____

_____

_____

UNDERSTANDING KEY IDEAS

_____ 2. Which of the following affects climate by causing the air to rise?
   a. mountains
   b. ocean currents
   c. large bodies of water
   d. latitude

3. What is the difference between weather and climate?

_____

_____

_____

_____

4. List five factors that determine climates.

_____

_____

_____

_____

_____

Copyright © by Holt, Rinehart and Winston. All rights reserved.

208

5. Explain why there is a difference in climate between areas at 0° latitude and areas at 45° latitude.

_____

_____

_____

_____

6. List the three climate zones of the world.

_____

_____

_____

CRITICAL THINKING

7. Analyzing Relationships How would seasons be different if the Earth did not tilt on its axis?

_____

_____

_____

8. Applying Concepts During what months does Australia have summer? Explain.

_____

_____

_____

_____

_____

INTERPRETING GRAPHICS

Use the map below to answer the questions that follow.

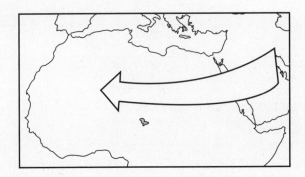

9. Would you expect the area that the arrow points to to be moist or dry? Explain your answer.

_____

_____

_____

_____

10. Describe how the climate of the same area would change if the prevailing winds traveled from the opposite direction. Explain how you came to this conclusion.

_____

_____

_____

_____

Skills Worksheet

Section Review

The Tropics
USING KEY TERMS

1. In your own words, write a definition for the term *tropical zone.*

_____

_____

UNDERSTANDING KEY IDEAS

_____ **2.** Which of the following tropical biomes has less than 50 cm of precipitation a year?
   a. rain forest
   b. desert
   c. grassland
   d. savanna

3. What are the soil characteristics of a tropical rain forest?

_____

_____

_____

4. In what ways have savanna vegetation adapted to fire?

_____

_____

_____

MATH SKILLS

5. Suppose that in a tropical savanna, the temperature was recorded every hour for 4 h. The recorded temperatures were 27°C, 28°C, 29°C, and 29°C. Calculate the average temperature for this 4 h period. Show your work below.

| Section Review *continued*

CRITICAL THINKING

6. Analyzing Relationships How do the tropical biomes differ?

_____

_____

_____

_____

7. Making Inferences How would you expect the adaptations of a plant in a tropical rain forest to differ from the adaptations of a tropical desert plant? Explain.

_____

_____

_____

_____

8. Analyzing Data An area has a temperature range of 30°C to 40°C and received 10 cm of rain this year. What biome is this area in?

_____

_____

_____

_____

Skills Worksheet

Section Review

Temperate and Polar Zones
USING KEY TERMS

1. In your own words, write a definition for the term *microclimate*.

_____

_____

Complete each of the following sentences by choosing the correct term from the word bank.

    temperate zone   polar zone   microclimate

2. The coldest temperatures are found in the _____ .

3. The _____ has moderate temperatures.

UNDERSTANDING KEY IDEAS

_____ **4.** Which of the following biomes has the driest climate?
    a. temperate forests
    b. temperate grasslands
    c. chaparrals
    d. temperate deserts

5. Explain why the temperate zone has lower temperatures than the Tropics.

_____

_____

6. Describe how the latitude of the polar zone affects the climate in that area.

_____

_____

7. Explain why the tundra can sometimes experience 24 hours of daylight or 24 hours of night.

_____

_____

_____

_____

Section Review *continued*

8. How do conifers make the soil they grow in too acidic for other plants to grow?

_____

_____

_____

MATH SKILLS

9. Texas has an area of about 700,000 square kilometers. Grasslands compose about 20% of this area. About how many square kilometers of grassland are there in Texas? Show your work below.

CRITICAL THINKING

10. Identifying Relationships Which biome would be more suitable for growing crops, temperate forest or taiga? Explain.

_____

_____

_____

_____

_____

11. Making Inferences Describe the types of animals and vegetation you might find in the Alpine biome.

_____

_____

_____

_____

Skills Worksheet

Section Review

Changes in Climate

USING KEY TERMS

1. Use the following term in a sentence: *ice age.*

_____

_____

2. In your own words, write a definition for each of the following terms: *global warming* and *greenhouse effect.*

_____

_____

UNDERSTANDING KEY IDEAS

3. Describe the possible causes of an ice age.

_____

_____

_____

_____ **4.** Which of the following can cause a change in the climate due to dust particles?

  a. volcanic eruptions

  b. plate tectonics

  c. solar cycles

  d. ice ages

5. How has the Earth's climate changed over time?

_____

_____

6. What might have caused the Earth's climate to change?

_____

_____

7. Which period of an ice age are we in currently? Explain.

_____

_____

| **Section Review** *continued* | | Climate |

8. Explain how the greenhouse effect warms the Earth.

_____

_____

MATH SKILLS

9. After a volcanic eruption, the average temperature in a region dropped from 30° to 18°C. By how many degrees Celsius did the temperature drop? Show your work below.

CRITICAL THINKING

10. Analyzing Relationships How will the warming of the Earth affect agriculture in different parts of the world? Explain.

_____

_____

_____

_____

11. Predicting Consequences How would deforestation (the cutting of trees) affect global warming?

_____

_____

_____

_____

Skills Worksheet

Chapter Review

USING KEY TERMS

For each pair of terms, explain how the meanings of the terms differ.

1. *biome* and *tropical zone*

_____

_____

2. *weather* and *climate*

_____

_____

3. *temperate zone* and *polar zone*

_____

_____

Complete each of the following sentences by choosing the correct term from the word bank.

> biome   microclimate
> ice age   global warming

4. One factor that could add to _____ is an increase in pollution.

5. A city is an example of a(n) _____.

UNDERSTANDING KEY IDEAS

Multiple Choice

_____ 6. Which of the following is a factor that affects climate?
   a. prevailing winds   **c.** ocean currents
   b. latitude   **d.** All of the above

_____ 7. The biome that has a temperature range of 28°C to 32°C and an average yearly precipitation of 100 cm is the
   a. tropical savanna.   **c.** tropical rain forest.
   b. tropical desert.   **d.** None of the above

_____ 8. Which of the following biomes is NOT found in the temperate zone?
   a. temperate forest   **c.** chaparral
   b. taiga   **d.** temperate grassland

_____ **9.** In which of the following is the tilt of the Earth's axis considered to have an effect on climate?

    a. global warming

    b. the sun's cycle

    c. the Milankovitch theory

    d. asteroid impact

_____ **10.** Which of the following substances contributes to the greenhouse effect?

    a. smoke

    b. smog

    c. carbon dioxide

    d. All of the above

_____ **11.** In which of the following climate zones is the soil most fertile?

    a. the tropical climate zone

    b. the temperate climate zone

    c. the polar climate zone

    d. None of the above

Short Answer

12. Why do higher latitudes receive less solar radiation than lower latitudes do?

_____

_____

_____

13. How does wind influence precipitation patterns?

_____

_____

_____

14. Give an example of a microclimate. What causes the unique temperature and precipitation characteristics of this area?

_____

_____

_____

15. How are tundras and deserts similar?

_____

_____

16. How does deforestation influence global warming?

_____

_____

_____

CRITICAL THINKING

17. Concept Mapping Use the following terms to create a concept map: *global warming, deforestation, changes in climate, greenhouse effect, ice ages,* and *the Milankovitch theory.*

18. Analyzing Processes Explain how ocean surface currents cause milder climates.

_____

_____

_____

19. Identifying Relationships Describe how the tilt of the Earth's axis affects seasonal changes in different latitudes.

_____

_____

_____

20. Evaluating Conclusions Explain why the climate on the eastern side of the Rocky Mountains differs drastically from the climate on the western side.

_____

_____

_____

21. Applying Concepts What are some steps you and your family can take to reduce the amount of carbon dioxide that is released into the atmosphere?

_____

_____

_____

22. Applying Concepts If you wanted to live in a warm, dry area, which biome would you choose to live in?

_____

_____

_____

23. Evaluating Data Explain why the vegetation in areas that have a tundra climate is sparse even though these areas receive precipitation that is adequate to support life.

_____

_____

_____

INTERPRETING GRAPHICS

Use the diagram below to answer the questions that follow.

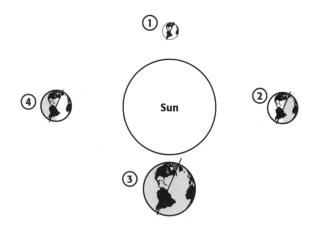

24. At what position—1, 2, 3, or 4—is it spring in the Southern Hemisphere?

_____

25. At what position does the South Pole receive almost 24 hours of daylight?

_____

26. Explain what is happening in each climate zone in both the Northern and Southern Hemispheres at position 4.

_____

_____

_____

_____

_____

_____

Skills Worksheet

Section Review

Astronomy: The Original Science

USING KEY TERMS

1. Use each of the following terms in a separate sentence: *year, day, month,* and *astronomy.*

_____

_____

_____

_____

_____

_____

_____

_____

UNDERSTANDING KEY IDEAS

_____ **2.** What happens in 1 year?
   a. The moon completes one orbit around the Earth.
   b. The sun travels once around the Earth.
   c. The Earth revolves once on its axis.
   d. The Earth completes one orbit around the sun.

3. What is the difference between the Ptolemaic and Copernican theories? Who was more accurate: Ptolemy or Copernicus?

_____

_____

_____

_____

4. What contributions did Brahe and Kepler make to astronomy?

_____

_____

_____

_____

5. What contributions did Galileo, Newton, and Hubble make to astronomy?

_____

_____

_____

_____

MATH SKILLS

6. How many times did Earth orbit the sun between 140 CE, when Ptolemy introduced his theories, and 1543, when Copernicus introduced his theories? Show your work below.

CRITICAL THINKING

7. Analyzing Relationships What advantage did Galileo have over earlier astronomers?

_____

_____

_____

_____

8. Making Inferences Why is astronomy such an old science?

_____

_____

_____

_____

Skills Worksheet)

Section Review

Telescopes

USING KEY TERMS

For each pair of terms, explain how the meanings of the terms differ.

1. *refracting telescope* and *reflecting telescope*

_____

_____

2. *telescope* and *electromagnetic spectrum*

_____

_____

UNDERSTANDING KEY IDEAS

_____ **3.** How does the atmosphere affect astronomical observations?
  a. It focuses visible light.
  b. It blocks most electromagnetic radiation.
  c. It blocks all radio waves.
  d. It does not affect astronomical observations.

4. Describe how reflecting and refracting telescopes work.

_____

_____

_____

_____

5. What limits the size of a refracting telescope? Explain.

_____

_____

6. What advantages do reflecting telescopes have over refracting telescopes?

_____

_____

_____

_____

7. List the types of radiation in the electromagnetic spectrum, from the longest wavelength to the shortest wavelength. Then, describe how astronomers study each type of radiation.

_____

_____

_____

MATH SKILLS

8. A telescope's light-gathering power is proportional to the area of its objective lens or mirror. If the diameter of a lens is 1 m, what is the area of the lens? (Hint: $area = 3.1416 \times radius^2$) Show your work below.

CRITICAL THINKING

9. Applying Concepts Describe three reasons why Hawaii is a good location for a telescope.

_____

_____

_____

10. Making Inferences Why doesn't the surface of a radio telescope have to be as flawless as the surface of a mirror in an optical telescope?

_____

_____

_____

11. Making Inferences What limitation of a refracting telescope could be overcome by placing the telescope in space?

_____

_____

_____

Section Review

Mapping the Stars
USING KEY TERMS

The statements below are false. For each statement, replace the underlined term to make a true statement.

1. Zenith is the angle between an object and the horizon.

_____

_____

2. The distance that light travels in 1 year is called a light-meter.

_____

_____

UNDERSTANDING KEY IDEAS

_____ 3. Stars appear to move across the sky during the night because of
  a. the rotation of Earth on its axis.
  b. the movement of the Milky Way galaxy.
  c. the movement of stars in the universe.
  d. the revolution of Earth around the sun.

4. How do astronomers use the celestial sphere to plot a star's exact position?

_____

_____

5. How do constellations relate to patterns of stars? How are constellations like states?

_____

_____

6. Why are different sky maps needed for different times of the year?

_____

_____

7. What are redshift and blueshift? Why are these effects useful in the study of the universe?

_____

_____

Section Review *continued*

CRITICAL THINKING

8. Applying Concepts Light from the Andromeda galaxy is affected by blueshift. What can you conclude about this galaxy?

_____

_____

9. Making Comparisons Explain how Copernicus concluded that stars were farther away than planets. Draw a diagram showing how this principle applies to another example.

INTERPRETING GRAPHICS

The diagram below shows the altitude of Star A and Star B. Use the diagram below to answer the questions that follow.

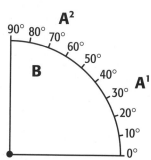

10. What is the approximate altitude of Star B?

_____

_____

11. In 4 h, Star A moved from A^1 to A^2. How many degrees did the star move each hour?

_____

_____

Skills Worksheet

Chapter Review

USING KEY TERMS

1. Use each of the following terms in a separate sentence: *year, month, day, astronomy, electromagnetic spectrum, constellation,* and *altitude.*

_____

_____

_____

_____

_____

_____

_____

For each pair of terms, explain how the meanings of the terms differ.

2. *reflecting telescope* and *refracting telescope*

_____

_____

_____

_____

3. *zenith* and *horizon*

_____

_____

_____

_____

4. *year* and *light-year*

_____

_____

_____

_____

| Chapter Review *continued*

UNDERSTANDING KEY IDEAS

Multiple Choice

_____ **5.** Which of the following answer choices lists types of electromagnetic radiation from longest wavelength to shortest wavelength?
  a. radio waves, ultraviolet light, infrared light
  b. infrared light, microwaves, X rays
  c. X rays, ultraviolet light, gamma rays
  d. microwaves, infrared light, visible light

_____ **6.** The length of a day is based on the amount of time that
  a. Earth takes to orbit the sun one time.
  b. Earth takes to rotate once on its axis.
  c. the moon takes to orbit Earth one time.
  d. the moon takes to rotate once on its axis.

_____ **7.** Which of the following statements about X rays and radio waves from objects in space is true?
  a. Both types of radiation can be observed by using the same telescope.
  b. Separate telescopes are needed to observe each type of radiation, but both telescopes can be on Earth.
  c. Separate telescopes are needed to observe each type of radiation, but both telescopes must be in space.
  d. Separate telescopes are needed to observe each type of radiation, but only one of the telescopes must be in space.

_____ **8.** According to _____, Earth is at the center of the universe.
  a. the Ptolemaic theory
  b. Copernicus's theory
  c. Galileo's theory
  d. None of the above

_____ **9.** Which scientist was one of the first scientists to successfully use a telescope to observe the night sky?
  a. Brahe   **c.** Hubble
  b. Galileo   **d.** Kepler

_____ **10.** Astronomers divide the sky into
  a. galaxies.   **c.** zeniths.
  b. constellations. **d.** phases.

_____ **11.** _____ determines which stars you see in the sky.
  a. Your latitude
  b. The time of year
  c. The time of night
  d. All of the above

_____**12.** The altitude of an object in the sky is the object's angular distance
 a. above the horizon.
 b. from the north celestial pole.
 c. from the zenith.
 d. from the prime meridian.

_____**13.** Right ascension is a measure of how far east an object in the sky
 is from
 a. the observer.
 b. the vernal equinox.
 c. the moon.
 d. Venus.

_____**14.** Telescopes that work on Earth's surface include all of the following
 EXCEPT
 a. radio telescopes.
 b. refracting telescopes.
 c. X-ray telescopes.
 d. reflecting telescopes.

Short Answer

15. Explain how right ascension and declination are similar to latitude and
 longitude.

_____

_____

_____

_____

_____

16. How does a reflecting telescope work?

_____

_____

_____

_____

_____

CRITICAL THINKING

17. Concept Mapping Use the following terms to create a concept map: *right ascension, declination, celestial sphere, degrees, hours, celestial equator,* and *vernal equinox.*

18. Making Inferences Why was seeing objects in the sky easier for people in ancient cultures than it is for most people today? What tools help modern people study objects in space in greater detail than was possible in the past?

_____

_____

_____

_____

_____

_____

19. Making Inferences Because many forms of radiation from space do not penetrate Earth's atmosphere, astronomers' ability to detect this radiation is limited. But how does the protection of the atmosphere benefit humans?

_____

_____

_____

_____

_____

20. Analyzing Ideas Explain why the Ptolemaic theory seems logical based on daily observations of the rising and setting of the sun.

_____

_____

_____

INTERPRETING GRAPHICS

Use the sky map below to answer the questions that follow. (Example: The star Aldebaran is located at about 4 h, 30 min right ascension, 16° declination.)

Celestial Coordinates

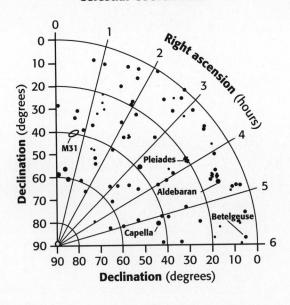

21. What object is located near 5 h, 55 min right ascension, and 7° declination?

_____

_____

_____

22. What are the celestial coordinates for the Andromeda galaxy (M31)? Round off the right ascension to the nearest half-hour.

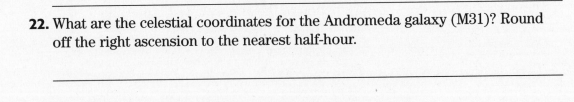

Skills Worksheet

Section Review

Stars

USING KEY TERMS

1. Use the following terms in the same sentence: *apparent magnitude* and *absolute magnitude*.

_____

_____

2. Use each of the following terms in a separate sentence: *spectrum*, *light-year*, and *parallax*.

_____

_____

_____

UNDERSTANDING KEY IDEAS

_____ **3.** When you look at white light through a glass prism, you see a rainbow of colors called a
    a. spectrograph.
    b. spectrum.
    c. parallax.
    d. light-year.

_____ **4.** Class F stars are
    a. blue.
    b. yellow.
    c. yellow-white.
    d. red.

5. Describe how scientists classify stars.

_____

_____

_____

6. Explain how color indicates the temperature of a star.

_____

_____

_____

CRITICAL THINKING

7. Applying Concepts If a certain star displayed a large parallax, what could you say about the star's distance from Earth?

_____

_____

_____

8. Making Comparisons Compare a continuous spectrum with an absorption spectrum. Then, explain how an absorption spectrum can identify a star's composition.

_____

_____

_____

9. Making Comparisons Compare apparent motion with actual motion.

_____

_____

_____

INTERPRETING GRAPHICS

10. Look at the two figures below. How many hours passed between the first image and the second image? Explain your answer.

_____

_____

_____

_____

Skills Worksheet

Section Review

The Life Cycle of Stars
USING KEY TERMS

For each pair of terms, explain how the meanings of the terms differ.

1. *white dwarf* and *red giant*

_____

_____

2. *supernova* and *neutron star*

_____

_____

3. *pulsar* and *black hole*

_____

_____

UNDERSTANDING KEY IDEAS

_____ **4.** The sun is a
    a. white dwarf.   **c.** red giant.
    b. main-sequence star.   **d.** red dwarf.

_____ **5.** A star begins as a ball of gas and dust pulled together by
    a. black holes.   **c.** heavy metals.
    b. electrons and protons.   **d.** gravity.

6. Are blue stars young or old? How can you tell?

_____

_____

7. In main-sequence stars, what is the relationship between brightness and temperature?

_____

_____

| Section Review *continued*

8. Arrange the following stages in order of their appearance in the life cycle of a star: white dwarf, red giant, and main-sequence star. Explain your answer.

_____

_____

_____

MATH SKILLS

9. The sun's present radius is 700,000 km. If the sun's radius increased by 150 times, what would its radius be? Show your work below.

CRITICAL THINKING

10. **Applying Concepts** Given that there are more low-mass stars than high-mass stars in the universe, do you think there are more white dwarfs or more black holes in the universe? Explain.

_____

_____

_____

_____

11. **Analyzing Processes** Describe what might happen to a star after it becomes a supernova.

_____

_____

_____

12. **Evaluating Data** How does the H-R diagram explain the life cycle of a star?

_____

_____

_____

Skills Worksheet

Section Review

Galaxies

USING KEY TERMS

1. Use the following terms in the same sentence: *nebula*, *globular cluster*, and *open cluster*.

_____

_____

_____

UNDERSTANDING KEY IDEAS

2. Arrange the following galaxies in order of decreasing size: spiral, giant elliptical, dwarf elliptical, and irregular.

_____

_____

_____ **3.** All of the following are shapes used to classify galaxies EXCEPT
      a. elliptical.
      b. irregular.
      c. spiral.
      d. triangular.

CRITICAL THINKING

4. Making Comparisons Describe the difference between an elliptical galaxy and a globular cluster.

_____

_____

_____

5. Identifying Relationships Explain how looking through a telescope is like looking back in time.

_____

_____

_____

| Section Review *continued*

MATH SKILLS

6. The quasar known as PKS 0637-752 is 6 billion light-years away from Earth. The North Star is 431 light-years away from Earth. What is the ratio of the distances in kilometers between these two celestial objects are from Earth? (Hint: One light-year is equal to 9.46 trillion kilometers.) Show your work below.

Section Review

Formation of the Universe
USING KEY TERMS

1. In your own words, write a definition for the following terms: *cosmology* and *big bang theory*.

_____

_____

UNDERSTANDING KEY IDEAS

2. Describe two ways scientists calculate the age of the universe.

_____

_____

_____ 3. The expansion of the universe can be compared to
   a. cosmology.
   b. raisin bread baking in an oven.
   c. thermal energy leaving an oven as the oven cools.
   d. bread pudding.

4. How does cosmic background radiation support the big bang theory?

_____

_____

_____

5. What do scientists think will eventually happen to the universe?

_____

_____

_____

| **Section Review** *continued*

MATH SKILLS

6. The North Star is 4.08×10^{12} km from Earth. What is this number written in its long form? Show your work below.

CRITICAL THINKING

7. Applying Concepts Explain how every object in the universe is part of a larger system.

_____

_____

_____

_____

_____

_____

8. Analyzing Ideas Why do scientists think that the universe will expand forever?

_____

_____

_____

_____

_____

_____

Chapter Review

USING KEY TERMS

The statements below are false. For each statement, replace the underlined term to make a true statement.

1. The distance that light travels in space in 1 year is called <u>apparent magnitude</u>.

2. <u>Globular clusters</u> are groups of stars that are usually located along the spiral disk of a galaxy.

3. Galaxies that have very bright centers and very little dust and gas are called <u>spiral galaxies</u>.

4. When you look at white light through a glass prism, you see a rainbow of colors called a <u>supernova</u>.

UNDERSTANDING KEY IDEAS

Multiple Choice

_____ **5.** A scientist can identify a star's composition by looking at
  a. the star's prism.
  b. the star's continuous spectrum.
  c. the star's absorption spectrum.
  d. the star's color.

_____ **6.** If the universe expands forever,
  a. the universe will collapse.
  b. the universe will repeat itself.
  c. the universe will remain just as it is today.
  d. stars will age and die and the universe will become cold and dark.

_____ **7.** The majority of stars in our galaxy are
  a. blue stars.
  b. white dwarfs.
  c. main-sequence stars.
  d. red giants.

_____ **8.** Which of the following is used to measure the distance between objects in space?
  a. parallax
  b. magnitude
  c. zenith
  d. altitude

_____ **9.** Which of the following stars would be seen as the brightest star?
  a. Alcyone, which has an apparent magnitude of 3
  b. Alpheratz, which has an apparent magnitude of 2
  c. Deneb, which has an apparent magnitude of 1
  d. Rigel, which has an apparent magnitude of 0

| Chapter Review *continued*

Short Answer

10. Describe how scientists classify stars.

_____

_____

11. Describe the structure of the universe.

_____

_____

_____

12. Explain how stars at different stages in their life cycle appear on the H-R diagram.

_____

_____

_____

13. Explain the difference between the apparent motion and actual motion of stars.

_____

_____

_____

14. Describe how color indicates the temperature of a star.

_____

_____

_____

15. Describe two ways that scientists calculate the age of the universe.

_____

_____

_____

CRITICAL THINKING

16. Concept Mapping Use the following terms to create a concept map: *main-sequence star, nebula, red giant, white dwarf, neutron star,* and *black hole.*

17. Evaluating Conclusions While looking through a telescope, you see a galaxy that doesn't appear to contain any blue stars. What kind of galaxy is it most likely to be? Explain your answer.

_____

_____

_____

_____

18. Making Comparisons Explain the differences between main-sequence stars, giant stars, supergiant stars, and white dwarfs.

_____

_____

_____

_____

19. Evaluating Data Why do astronomers use absolute magnitudes to plot stars? Why don't astronomers use apparent magnitudes to plot stars?

_____

_____

_____

20. Evaluating Sources According to the big bang theory, how did the universe begin? What evidence supports this theory?

_____

_____

_____

_____

21. Evaluating Data If a certain star displayed a large parallax, what could you say about the star's distance from Earth?

_____

_____

_____

INTERPRETING GRAPHICS

The graph below shows Hubble's law, which relates how far galaxies are from Earth and how fast they are moving away from Earth. Use the graph below to answer the questions that follow.

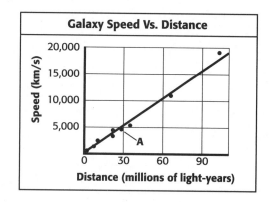

22. Look at the point that represents galaxy A in the graph. How far is galaxy A from Earth, and how fast is it moving away from Earth?

    _____

    _____

23. If a galaxy is moving away from Earth at 15,000 km/s, how far is the galaxy from Earth?

    _____

    _____

24. If a galaxy is 90,000,000 light-years from Earth, how fast is it moving away from Earth?

    _____

    _____

Section Review

A Solar System Is Born

USING KEY TERMS

1. In your own words, write a definition for each of the following terms: *nebula* and *solar nebula*.

_____

_____

UNDERSTANDING KEY IDEAS

_____ 2. What is the relationship between gravity and pressure in a nebula?
   a. Gravity reduces pressure.
   b. Pressure balances gravity.
   c. Pressure increases gravity.
   d. None of the above

3. Describe how our solar system formed.

_____

_____

_____

_____

4. Compare the inner planets with the outer planets.

_____

_____

_____

_____

MATH SKILLS

5. If the planets, moons, and other bodies make up 0.15% of the solar system's mass, what percentage does the sun make up? Show your work below.

CRITICAL THINKING

6. Evaluating Hypotheses Beyond the orbit of Neptune, smaller bodies orbit the sun. Some scientists think these smaller bodies are the remains of material that formed the early solar system. Use what you know about how solar systems form to evaluate this hypothesis..

_____

_____

_____

_____

7. Making Inferences Why do all of the planets go around the sun in the same direction, and why do the planets lie on a relatively flat plane?

_____

_____

_____

_____

Section Review

The Sun: Our Very Own Star

USING KEY TERMS

1. In your own words, write a definition for each of the following terms: *sunspot* and *nuclear fusion*.

_____

_____

UNDERSTANDING KEY IDEAS

_____ 2. Which of the following statements describes how energy is produced in the sun?
   a. The sun burns fuels to generate energy.
   b. As hydrogen changes into helium deep inside the sun, a great deal of energy is made.
   c. Energy is released as the sun shrinks because of gravity.
   d. None of the above

3. Describe the composition of the sun.

_____

_____

4. Name and describe the layers of the sun.

_____

_____

_____

5. In which area of the sun do sunspots appear?

_____

_____

6. Explain how sunspots form.

_____

_____

_____

7. Describe how sunspots can affect the Earth.

_____

_____

_____

8. What are solar flares, and how do they form?

_____

_____

_____

MATH SKILLS

9. If the equatorial diameter of the sun is 1.39 million km, how many kilometers is the sun's radius? Show your work below.

CRITICAL THINKING

10. Applying Concepts If nuclear fusion in the sun's core suddenly stopped today, would the sky be dark in the daytime tomorrow? Explain.

_____

_____

_____

_____

11. Making Comparisons Compare the theories that scientists proposed about the source of the sun's energy with the process of nuclear fusion in the sun.

_____

_____

_____

_____

Skills Worksheet

Section Review

The Earth Takes Shape

USING KEY TERMS

1. Use each of the following terms in a separate sentence: *crust, mantle,* and *core.*

_____

_____

_____

UNDERSTANDING KEY IDEAS

_____ 2. Earth's first atmosphere was mostly made of
    a. nitrogen and oxygen.
    b. chlorine, nitrogen, and sulfur.
    c. carbon dioxide and water vapor.
    d. water vapor and oxygen.

3. Describe the structure of the Earth.

_____

_____

_____

4. Why did the Earth separate into distinct layers?

_____

_____

_____

5. Describe the development of Earth's atmosphere. How did life affect Earth's atmosphere?

_____

_____

_____

_____

6. Explain how Earth's oceans and continents formed.

_____

_____

CRITICAL THINKING

7. Applying Concepts How did the effects of gravity help shape the Earth?

_____

_____

_____

8. Making Inferences How would the removal of forests affect the Earth's atmosphere?

_____

_____

_____

INTERPRETING GRAPHICS

Use the illustration below to answer the questions that follow.

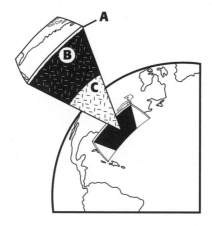

9. Which of the layers is composed mostly of the elements magnesium and iron?

_____

10. Which of the layers is composed mostly of the elements iron and nickel?

_____

Section Review

Planetary Motion

USING KEY TERMS

1. In your own words, write a definition for each of the following terms:
revolution and *rotation*.

_____

_____

_____

_____

UNDERSTANDING KEY IDEAS

_____ **2.** Kepler discovered that planets move faster when they
   a. are farther from the sun.
   b. are closer to the sun.
   c. have more mass.
   d. rotate faster.

3. On what properties does the force of gravity between two objects depend?

_____

_____

_____

_____

4. How does gravity keep a planet moving in an orbit around the sun?

_____

_____

_____

_____

| **Section Review** *continued*

MATH SKILLS

5. The Earth's period of revolution is 365.25 days. Convert this period of revolution into hours. Show your work below.

CRITICAL THINKING

6. Applying Concepts If a planet had two moons and one moon was twice as far from the planet as the other, which moon would complete a revolution of the planet first? Explain your answer.

_____

_____

_____

_____

_____

_____

7. Making Comparisons Describe the three laws of planetary motion. How is each law related to the other laws?

_____

_____

_____

_____

_____

_____

_____

Skills Worksheet)

Chapter Review

USING KEY TERMS

Complete each of the following sentences by choosing the correct term from the word bank.

nebula   crust

mantle   solar nebula

1. A _____ is a large cloud of gas and dust in interstellar space.

2. The _____ lies between the core and the crust of the Earth.

For each pair of terms, explain how the meanings of the terms differ.

3. *nebula* and *solar nebula*

_____

_____

4. *crust* and *mantle*

_____

_____

5. *rotation* and *revolution*

_____

_____

6. *nuclear fusion* and *sunspot*

_____

_____

UNDERSTANDING KEY IDEAS
Multiple Choice

_____ **7.** To determine a planet's period of revolution, you must know its
    a. size.
    b. mass.
    c. orbit.
    d. All of the above

Chapter Review *continued*

_____ **8.** During Earth's formation, materials such as nickel and iron sank to the
     a. mantle.   **c.** crust.
     b. core.   **d.** All of the above

_____ **9.** Planetary orbits are shaped like
     a. orbits.
     b. spirals.
     c. ellipses.
     d. periods of revolution.

_____ **10.** Impacts in the early solar system
     a. brought new materials to the planets.
     b. released energy.
     c. dug craters.
     d. All of the above

_____ **11.** Organisms that photosynthesize get their energy from
     a. nitrogen.   **c.** the sun.
     b. oxygen.   **d.** water.

_____ **12.** Which of the following planets has the shortest period of revolution?
     a. Mars   **c.** Mercury
     b. Earth   **d.** Jupiter

_____ **13.** Which gas in Earth's atmosphere suggests that there is life on Earth?
     a. hydrogen   **c.** carbon dioxide
     b. oxygen   **d.** nitrogen

_____ **14.** Which layer of the Earth has the lowest density?
     a. the core   **c.** the crust
     b. the mantle   **d.** None of the above

_____ **15.** What is the measure of the average kinetic energy of particles in an object?
     a. temperature   **c.** gravity
     b. pressure   **d.** force

Short Answer

16. Compare a sunspot with a solar flare.

_____

_____

17. Describe how the Earth's oceans and continents formed.

_____

_____

Chapter Review *continued*

18. Explain how pressure and gravity may have become unbalanced in the solar nebula.

_____

_____

19. Define *nuclear fusion* in your own words. Describe how nuclear fusion generates the sun's energy.

_____

_____

CRITICAL THINKING

20. Concept Mapping Use the following terms to create a concept map: *solar nebula, solar system, planetesimals, sun, photosphere, core, nuclear fusion, planets,* and *Earth.*

21. Making Comparisons How did Newton's law of universal gravitation help explain the work of Johannes Kepler?

_____

_____

_____

22. Predicting Consequences Using what you know about the relationship between living things and the development of Earth's atmosphere, explain how the formation of ozone holes in Earth's atmosphere could affect living things.

_____

_____

23. Identifying Relationships Describe Kepler's three laws of motion in your own words. Describe how each law relates to either the revolution, rotation, or orbit of a planetary body.

_____

_____

_____

INTERPRETING GRAPHICS

Use the illustration below to answer the questions that follow.

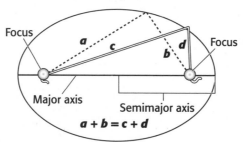

24. Which of Kepler's laws of motion does the illustration represent?

_____

25. How does the equation shown above support the law?

_____

_____

26. What is an ellipse's maximum length called?

_____

Skills Worksheet

Section Review

Our Solar System

USING KEY TERMS

1. In your own words, write a definition for the term *astronomical unit*.

_____

_____

_____

UNDERSTANDING KEY IDEAS

_____ 2. When was the planet Uranus discovered?
   a. before the 17th century
   b. in the 18th century
   c. in the 19th century
   d. in the 20th century

3. The invention of what instrument helped early scientists discover more bodies in the solar system?

_____

_____

4. Which of the eight planets are included in the outer solar system?

_____

_____

5. Describe how the inner planets are different from the outer planets.

_____

_____

_____

MATH SKILLS

6. If Venus is 6.0 light-minutes from the sun, what is Venus's distance from the sun in astronomical units? Show your work below.

Section Review *continued*

CRITICAL THINKING

7. Analyzing Methods The distance between Earth and the sun is measured in light-minutes, but the distance between Neptune and the sun is measured in light-hours. Explain why.

_____

_____

_____

_____

_____

Skills Worksheet

Section Review

The Inner Planets

USING KEY TERMS

1. In your own words, write a definition for the term *terrestrial planet*.

_____

_____

For the pair of terms below, explain how the meanings of the terms differ.

2. *prograde rotation* and *retrograde rotation*

_____

_____

UNDERSTANDING KEY IDEAS

_____ 3. Scientists believe that the water on Mars now exists as
   a. polar icecaps.
   b. dry riverbeds.
   c. ice beneath the Martian soil.
   d. Both (a) and (c)

4. List three differences between and three similarities of Venus and Earth.

_____

_____

_____

5. What is the difference between a planet's period of rotation and its period of revolution?

_____

_____

_____

6. What are some of the characteristics of Earth that make it suitable for life?

_____

_____

_____

7. Explain why the surface temperature of Venus is higher than the surface temperatures of the other planets in our solar system.

_____

_____

_____

MATH SKILLS

8. Mercury has a period of rotation equal to 58.67 Earth days. Mercury's period of revolution is equal to 88 Earth days. How many times does Mercury rotate during one revolution around the sun? Show your work below.

CRITICAL THINKING

9. **Making Inferences** What type of information can we get by studying Earth from space?

_____

_____

_____

_____

10. **Analyzing Ideas** What type of evidence found on Mars suggests that Mars may have been a warmer place and had a thicker atmosphere?

_____

_____

_____

_____

Skills Worksheet

Section Review

The Outer Planets
USING KEY TERMS

1. In your own words, write a definition for the term *gas giant*.

_____

_____

UNDERSTANDING KEY IDEAS

_____ **2.** The many colors of Jupiter's atmosphere are probably caused by _____ in the atmosphere.
  a. clouds of water
  b. methane
  c. ammonia
  d. organic compounds

3. Why do scientists claim that Saturn, in a way, is still forming?

_____

_____

4. Why does Uranus have a blue green color?

_____

_____

5. What is unusual about Pluto's moon, Charon?

_____

_____

6. What is the Great Red Spot?

_____

_____

7. Explain why Jupiter radiates more energy into space than it receives from the sun.

_____

_____

_____

8. How do the gas giants differ from the terrestrial planets?

_____

_____

9. What is so unusual about Uranus's axis of rotation?

_____

_____

MATH SKILLS

10. Pluto is 5.5 light-hours from the sun. How far is Pluto from the sun in astronomical units? (Hint: 1 AU = 8.3 light-minutes) Show your work below.

11. If Jupiter is 43.3 light-minutes from the sun and Neptune is 4.2 light-hours from the sun, how far from Jupiter is Neptune? Show your work below.

CRITICAL THINKING

12. Evaluating Data What conclusions can your draw about the properties of a planet just by knowing how far it is from the sun?

_____

_____

_____

13. Applying Concepts Why isn't the word *surface* included in the statistics for the gas giants?

_____

_____

_____

Section Review

Moons

USING KEY TERMS

Complete each of the following sentences by choosing the correct term from the word bank.

satellite   eclipse

1. A(n) _____, or a body that revolves around a larger body, can be either artificial or natural.

2. A(n) _____ occurs when the shadow of one body in space falls on another body.

UNDERSTANDING KEY IDEAS

_____ **3.** Which of the following is a Galilean satellite?
   a. Phobos
   b. Deimos
   c. Ganymede
   d. Charon

4. Describe the current theory for the origin of Earth's moon.

_____

_____

_____

5. What is the difference between a solar eclipse and a lunar eclipse?

_____

_____

_____

6. What causes the phases of Earth's moon?

_____

_____

_____

_____

CRITICAL THINKING

7. Analyzing Methods How can astronomers use the age of a lunar rock to estimate the age of the surface of a planet such as Mercury?

_____

_____

_____

8. Identifying Relationships Charon stays in the same place in Pluto's sky, but the moon moves across Earth's sky. What causes this difference?

_____

_____

_____

INTERPRETING GRAPHICS

Use the diagram below to answer the questions that follow.

9. What type of eclipse is shown in the diagram?

_____

10. Describe what is happening in the diagram.

_____

_____

11. Make a sketch of the type of eclipse that is not shown in the diagram.

Section Review

Small Bodies in the Solar System

USING KEY TERMS

For each pair of terms, explain how the meanings of the terms differ.

1. *comet* and *asteroid*

_____

_____

2. *meteor* and *meteorite*

_____

_____

_____

UNDERSTANDING KEY IDEAS

_____ **3.** Which of the following is NOT a type of meteorite?
   a. stony meteorite
   b. rocky-iron meteorite
   c. stony-iron meteorite
   d. metallic meteorite

4. Why is the study of comets, asteroids, and meteoroids important in understanding the formation of the solar system?

_____

_____

_____

5. Why do a comet's two tails often point in different directions?

_____

_____

6. How can a cosmic impact affect life on Earth?

_____

_____

7. What is the difference between an asteroid and a meteoroid?

_____

_____

8. Where is the asteroid belt located?

_____

_____

9. What is the Torino scale?

_____

_____

10. Describe why we see several impact craters on the moon but few on Earth.

_____

_____

MATH SKILLS

11. The diameter of comet A's nucleus is 55 km. If the diameter of comet B's nucleus is 30% larger than comet A's nucleus, what is the diameter of comet B's nucleus? Show your work below.

CRITICAL THINKING

12. Expressing Opinions Do you think the government should spend money on programs to search for asteroids and comets that have Earth-crossing orbits? Explain.

_____

_____

_____

13. Making Inferences What is the likelihood that scientists will discover an object belonging in the red category of the Torino scale in the next 500 years? Explain your answer.

_____

_____

_____

Skills Worksheet

Chapter Review

USING KEY TERMS

For each pair of terms, explain how the meanings of the terms differ.

1. *terrestrial planet* and *gas giant*

_____

_____

2. *asteroid* and *comet*

_____

_____

3. *meteor* and *meteorite*

_____

_____

Complete each of the following sentences by choosing the correct term from the word bank.

| astronomical unit | meteorite | meteoroid |
| prograde | retrograde | satellite |

4. The average distance between the sun and Earth is 1

_____.

5. A small rock in space is called a(n) _____.

6. When viewed from above its north pole, a body that moves in a counter-

clockwise direction is said to have _____ rotation.

7. A(n) _____ is a natural or artificial body that revolves

around a planet.

UNDERSTANDING KEY IDEAS

Multiple Choice

_____ 8. Of the following, which is the largest body?
   a. the moon
   b. Pluto
   c. Mercury
   d. Ganymede

Chapter Review *continued*

_____ **9.** Which of the following planets have retrograde rotation?
  a. the terrestrial planets   **c.** Mercury and Venus
  b. the gas giants   **d.** Venus and Uranus

_____ **10.** Which of the following planets does NOT have any moons?
  a. Mercury   **c.** Uranus
  b. Mars   **d.** None of the above

_____ **11.** Why can liquid water NOT exist on the surface of Mars?
  a. The temperature is too high.
  b. Liquid water once existed there.
  c. The gravity of Mars is too weak.
  d. The atmospheric pressure is too low.

Short Answer

12. List the names of the planets in the order the planets orbit the sun.

_____

_____

13. Describe three ways in which the inner planets are different from the outer planets.

_____

_____

_____

14. What are the gas giants? How are the gas giants different from the terrestrial planets?

_____

_____

15. What is the difference between asteroids and meteoroids?

_____

16. What is the difference between a planet's period of rotation and period of revolution?

_____

_____

17. Explain the difference between prograde rotation and retrograde rotation.

_____

_____

18. Which characteristics of Earth make it suitable for life?

_____

_____

19. Describe the current theory for the origin of Earth's moon.

_____

_____

20. What causes the phases of the moon?

_____

_____

CRITICAL THINKING

21. Concept Mapping Use the following terms to create a concept map:
solar system, terrestrial planets, gas giants, moons, comets, asteroids, and
meteoroids.

22. Applying Concepts Even though we haven't yet retrieved any rock samples from Mercury's surface for radiometric dating, scientists know that the surface of Mercury is much older than that of Earth. How do scientists know this?

_____

_____

_____

23. Making Inferences Where in the solar system might scientists search for life, and why?

_____

_____

_____

_____

24. Analyzing Ideas Is the far side of the moon always dark? Explain your answer.

_____

_____

_____

_____

25. Predicting Consequences If scientists could somehow bring Europa as close to the sun as the Earth is, 1 AU, how do you think Europa would be affected?

_____

_____

_____

26. Identifying Relationships How did variations in the orbit of Uranus help scientists discover Neptune?

_____

_____

_____

❙ Chapter Review *continued*

INTERPRETING GRAPHICS

The graph below shows density versus mass for Earth, Uranus, and Neptune.
Mass is given in Earth masses—the mass of Earth is equal to 1 Earth mass. The
relative volumes for the planets are shown by the size of each circle. Use the
graph below to answer the questions that follow.

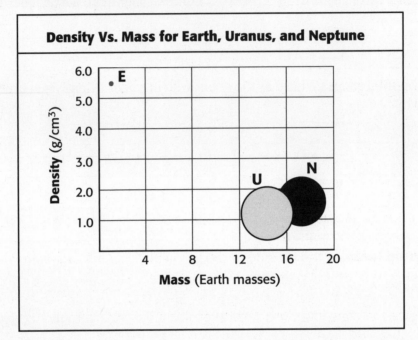

27. Which planet is denser, Uranus or Neptune? How can you tell?

_____

_____

_____

28. You can see that although Earth has the smallest mass, it has the highest
density of the three planets. How can Earth be the densest of the three when
Uranus and Neptune have so much more mass than Earth does?

_____

_____

_____

Skills Worksheet

Section Review

Rocket Science
USING KEY TERMS

1. Use each of the following terms in a separate sentence: *rocket, thrust,* and *NASA*.

_____

_____

_____

_____

UNDERSTANDING KEY IDEAS

_____ **2.** What factor must a rocket overcome to reach escape velocity?
  a. Earth's axial tilt
  b. Earth's gravity
  c. the thrust of its engines
  d. Newton's third law of motion

3. Describe the contributions of Tsiolkovsky and Goddard to modern rocketry.

_____

_____

4. Use Newton's third law of motion to describe how rockets work.

_____

_____

_____

5. What is the difference between orbital and escape velocity?

_____

_____

6. How did the Cold War accelerate the U.S. space program?

_____

_____

_____

MATH SKILLS

7. If you travel at 60 mi/h, it takes about 165 days to reach the moon. Approximately how far away is the moon? Show your work below.

CRITICAL THINKING

8. Applying Concepts How do rockets accelerate in space?

_____

_____

_____

_____

_____

9. Making Inferences Why does escape velocity vary depending on the planet from which a rocket is launched?

_____

_____

_____

_____

_____

Skills Worksheet

Section Review

Artificial Satellites

USING KEY TERMS

1. Use each of the following terms in a separate sentence: *artificial satellite*, *low Earth orbit*, and *geostationary orbit*.

_____

_____

_____

UNDERSTANDING KEY IDEAS

_____ **2.** In a low Earth orbit, the speed of a satellite is
   a. slower than the rotational speed of the Earth.
   b. equal to the rotational speed of the Earth.
   c. faster than the rotational speed of the Earth.
   d. None of the above

3. What was the name of the first satellite placed in orbit?

_____

_____

4. List three ways that satellites benefit human society.

_____

_____

_____

5. What was the *Explorer 1*?

_____

_____

6. Explain the differences between LEO and GEO satellites.

_____

_____

7. How does the Global Positioning System work?

_____

_____

Section Review *continued*

8. How do communications satellites relay signals around the curved surface of Earth?

_____

_____

MATH SKILLS

9. The speed required to reach Earth orbit is 8 km/s. What does this equal in *meters per hour*? Show your work below.

CRITICAL THINKING

10. Applying Concepts The *Hubble Space Telescope* is located in LEO. Will the telescope move faster or slower around the Earth compared with a geostationary weather satellite? Explain.

_____

_____

_____

11. Applying Concepts To triangulate your location on a map, you need to know your distance from three points. If you knew your distance from two points, how many possible places could you occupy?

_____

_____

_____

_____

Skills Worksheet

Section Review

Space Probes
USING KEY TERMS

The statements below are false. For each statement, replace the underlined term to make a true statement.

1. *Luna 1* discovered evidence of water on the moon.

_____

2. *Venera 9* helped map 98% of Venus's surface.

_____

3. *Stardust* uses ion propulsion to accelerate.

_____

UNDERSTANDING KEY IDEAS

_____ **4.** What is the significance of the discovery of evidence of water on the moon?
- **a.** Water is responsible for the formation of craters.
- **b.** Water was left by early space probes.
- **c.** Water could be used by future moon colonies.
- **d.** The existence of water proves that there is life on the moon.

5. Describe three discoveries that have been made by space probes.

_____

_____

_____

_____

6. How do missions to Venus, Mars, and Titan help us understand Earth's environment?

_____

_____

_____

_____

_____

MATH SKILLS

7. Traveling at the speed of light, signals from *Voyager 1* take about 12 h to reach Earth. The speed of light is about 299,793 km/s, how far away is the probe? Show your work below.

CRITICAL THINKING

8. Making Inferences Why did we need space probes to discover water channels on Mars and evidence of ice on Europa?

_____

_____

_____

_____

9. Expressing Opinions What are the advantages of the new Discovery program over the older space-probe missions, and what are the disadvantages?

_____

_____

_____

_____

_____

10. Applying Concepts How does *Deep Space 1* use Newton's third law of motion to accelerate?

_____

_____

_____

_____

_____

Skills Worksheet

Section Review

People in Space

USING KEY TERMS

1. Use each of the following terms in a separate sentence: *space shuttle* and *space station*.

_____

_____

USING KEY IDEAS

_____ **2.** What is the main difference between the space shuttles and other space vehicles?

   a. The space shuttles are powered by liquid rocket fuel.
   b. The space shuttles take off like a plane and land like a rocket.
   c. The space shuttles are reusable.
   d. The space shuttles are not reusable.

3. Describe the history and future of human spaceflight. How was the race to explore space influenced by the Cold War?

_____

_____

_____

_____

4. Describe five "space-age spinoffs."

_____

_____

_____

_____

5. How will space stations help in the exploration of space?

_____

_____

6. In the 1970s, what was the main difference in the focus of the space programs in the United States and in the Soviet Union?

_____

_____

MATH SKILLS

7. When it is fueled, a space shuttle has a mass of about 2,000,000 kg. About 80% of that mass is fuel and oxygen. Calculate the weight of a space shuttle's fuel and oxygen. Show your work below.

CRITICAL THINKING

8. Making Inferences Why did the United States stop sending people to the moon after the Apollo program ended?

_____

_____

_____

_____

9. Expressing Opinions Imagine that you are a U.S. senator reviewing NASA's proposed budget. Write a two-paragraph position statement expressing your opinion about increasing or decreasing funding for NASA.

_____

_____

_____

_____

_____

Skills Worksheet

Chapter Review

USING KEY TERMS

For each pair of terms, explain how the meanings of the terms differ.

1. *geostationary orbit* and *low Earth orbit*

_____

_____

2. *space probe* and *space station*

_____

_____

3. *artificial satellite* and *moon*

_____

_____

_____

Complete each of the following sentences by choosing the correct term from the word bank.

| escape velocity | oxygen |
| nitrogen | thrust |

4. The force that accelerates a rocket is called _____.

5. Rockets need to have _____ in order to burn fuel.

UNDERSTANDING KEY IDEAS

Multiple Choice

_____ 6. Whose rocket research team surrendered to the Americans at the end of World War II?
   a. Konstantin Tsiolkovsky's   **c.** Wernher von Braun's
   b. Robert Goddard's   **d.** Yuri Gargarin's

_____ 7. Rockets work according to Newton's
   a. first law of motion.
   b. second law of motion.
   c. third law of motion.
   d. law of universal gravitation.

_____ 8. The first artificial satellite to orbit the Earth was
   a. *Pioneer 4.*   **c.** *Voyager 2.*
   b. *Explorer 1.*   **d.** *Sputnik 1.*

_____ **9.** Communications satellites are able to transfer TV signals between continents because communications satellites
 a. are located in LEO.
 b. relay signals past the horizon.
 c. travel quickly around Earth.
 d. can be used during the day and night.

_____ **10.** GEO is a better orbit for communications satellites because satellites that are in GEO
 a. remain in position over one spot.
 b. have polar orbits.
 c. do not revolve around the Earth.
 d. orbit a few hundred kilometers above the Earth.

_____ **11.** Which space probe discovered evidence of water at the moon's south pole?
 a. *Luna 9*   **c.** *Clementine*
 b. *Viking 1*   **d.** *Magellan*

_____ **12.** When did humans first set foot on the moon?
 a. 1959   **c.** 1969
 b. 1964   **d.** 1973

_____ **13.** Which of the following bodies has not yet been visited by space probes?
 a. Venus   **c.** Mars
 b. Neptune   **d.** Pluto

_____ **14.** Which of the following space probes has left our solar system?
 a. *Galileo*   **c.** *Viking 10*
 b. *Magellan*   **d.** *Pioneer 10*

_____ **15.** Based on space-probe data, which of the following is the most likely place in our solar system to find liquid water?
 a. the moon   **c.** Europa
 b. Mercury   **d.** Venus

Short Answer

16. Describe how Newton's third law of motion relates to the movement of rockets.

_____

_____

_____

17. What is one disadvantage that objects in LEO have?

_____

_____

18. Why did the United States develop the space shuttle?

_____

_____

19. How does data from satellites help us understand the Earth's environment?

_____

_____

20. Concept Mapping Use the following terms to create a concept map: *orbital velocity, thrust, LEO, artificial satellites, escape velocity, space probes, GEO,* and *rockets.*

Making Inferences What is the difference between speed and velocity?

_____

_____

_____

_____

_____

_____

22. Applying Concepts Why must rockets that travel in outer space carry oxygen with them?

_____

_____

_____

_____

_____

_____

23. Expressing Opinions What impact has space research had on scientific thought, on society, and on the environment?

_____

_____

_____

_____

_____

_____

The diagram below illustrates suborbital velocity, orbital velocity, and escape velocity. Use the diagram below to answer the questions that follow.

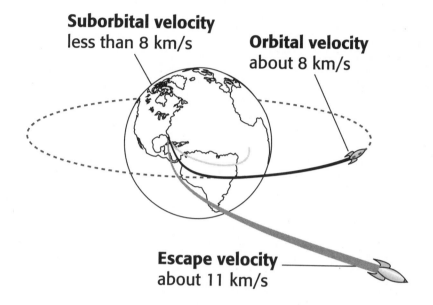

Suborbital velocity
less than 8 km/s

Orbital velocity
about 8 km/s

Escape velocity
about 11 km/s

24. Could a rocket traveling at 6 km/s reach orbital velocity?

_____

25. If a rocket traveled for 3 days at the minimum escape velocity, how far would the rocket travel?

_____

26. How much faster would a rocket traveling in orbital velocity need to travel to reach escape velocity?

_____

27. If the escape velocity for a planet was 9 km/s, would you assume that the mass of the planet was more or less than the mass of Earth?

_____